Mechatronics

Mechatronics

Edited by
J. Paolo Davim

First published 2011 in Great Britain and the United States by ISTE Ltd and John Wiley & Sons, Inc.

ISTE Ltd
27-37 St George's Road
London SW19 4EU
UK

www.iste.co.uk

John Wiley & Sons, Inc.
111 River Street
Hoboken, NJ 07030
USA

www.wiley.com

Library of Congress Cataloging-in-Publication Data

Mechatronics / edited by J. Paulo Davim.
    p. cm.
  Includes bibliographical references and index.
  ISBN 978-1-84821-308-1
  1. Mechatronics. I. Davim, J. Paulo.
  TJ163.12.M4224 2011
  621--dc22
                                                                                    2011006659

British Library Cataloguing-in-Publication Data
A CIP record for this book is available from the British Library
ISBN 978-1-84821-308-1

Printed and bound in Great Britain by CPI Antony Rowe, Chippenham and Eastbourne.

Table of Contents

Preface

The term Mechatronics is a portmanteau of "Mechanics" and "Electronics" (MECHAnics elecTRONICS). Mechatronics is the blending of mechanical, electronic, and computer engineering into an integrated design. Mechatronics systems use microprocessors and software, as well as special-purpose electronics. The main objective of this interdisciplinary engineering field is the study of automata from an engineering perspective, thinking of the design of products and manufacturing processes.

Today, mechatronics has a significant and increasing impact on engineering; on the design, development, and operation of engineering systems. Mechatronics systems and products are well established in a large number of industries such as aircraft, automotive, computers, electronics, robotics/automation, computerized machine tools, communications, and biomedical.

The aim of this book is to present a collection of examples illustrating the state-of-the-art and research developments in mechatronics. Chapter 1 presents mechatronics systems based on CAD/CAM. Chapter 2 covers modeling and control of ionic polymer–metal composite (IPMC) actuators for mechatronics applications. Chapter 3 covers modeling and simulation of analog angular sensors for manufacturing purposes. Chapter 4 contains information on robust control of atomic force microscopy. Chapter 5 is dedicated to automated identification. Chapter 6 covers an active orthosis for gait rehabilitation. Finally, in Chapter 7, an intelligent assistive knee exoskeleton is presented.

This book can be used as a reference for a final undergraduate engineering course or as a comprehensive study on mechatronics at the postgraduate level. Also, this book can serve as a useful reference for academics, mechatronics, and automation researchers; mechanical, manufacturing, and computer engineers; and professionals in mechatronics, robotics, and related industries. The interest of this book is evident for many important centers of the research, laboratories, and universities throughout the world. Therefore, it is hoped that this book will encourage and enthuse other research into this important field of engineering and technology.

The Editor acknowledges gratitude to ISTE-Wiley for this opportunity and for their professional support. Finally, I would like to thank all the chapter authors for their availability to work on this project.

J. Paulo Davim,
University of Aveiro, Portugal,
March 2011

Chapter 1

Mechatronics Systems Based on CAD/CAM

This chapter describes several mechatronics systems that use computer-aided design/computer-aided manufacturing (CAD/CAM) software. These mechatronics systems include: a five-axis NC machine tool with a tilting head, a three-axis NC machine tool with a rotary unit, articulated-type industrial robots with six degrees-of-freedom (DOF), and a desktop Cartesian-type robot with high-position resolution.

1.1. Introduction

Computer-aided design (CAD) involves creating computer models defined by geometrical parameters. Computer-aided manufacturing (CAM) uses geometrical design data to control automated machinery. CAD/CAM use computer-based tools and simulation that greatly assist the integration of design and manufacture processes. The main-processor of CAM generates cutter location data called CL data that contain precise position and orientation components, when the linear approximation mode is selected in CAM parameters. Hence, to achieve efficient implementation, machine tools and robotics can display superior abilities and high-level functions using the generated CL data.

1.2. Five-axis NC machine tool with a tilting head

In this section, a three-dimensional (3D) machining system based on a five-axis NC machine tool with a tilting head is evaluated as a conventional system

Chapter written by Fusaomi NAGATA, Yukihiro KUSUMOTO, Keigo WATANABE and Maki K. HABIB.

[NAG 96]. The five-axis NC machine tool is one of the most representative and popular machine tools in wood product manufacturing. The machining system consists of a 3D CAD/CAM with variable-axis function and a post-processor. Figure 1.1 shows the five-axis NC machine tool (HEIAN FF-151MC). The NC machine tool has a tilting head that can simultaneously incline and rotate. It should be noted, however, that corrected NC data are required to run the NC machine tool as a computer simulation. The post-process generally involves computing corrected NC data from CL data. Figure 1.2 shows the general process to calculate the corrected NC data for the five-axis NC machine tool.

Figure 1.1. *Five-axis NC machine tool with a tilting head*

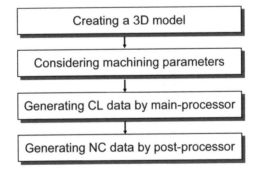

Figure 1.2. *Process to compute NC data for five-axis NC machine tool*

Examples of paint roller models with a relief design are shown in Figure 1.3. First, the model of a relief design was drawn using a 3D CAD. Second, CAM parameters such as pick feed, path pattern (e.g. zigzag path), in/out tolerances, and so on are set according to the requirement of actual machining. The main-processor of CAM calculates cutter paths using the parameters. The cutter paths are called CL data. The i-th step **CL**(i) in CL data is composed of position vector

$\mathbf{p}(i) = [x(i)\ y(i)\ z(i)]^{\mathrm{T}}$ and normal direction vector $\mathbf{n}(i) = [n_x(i)\ n_y(i)\ n_z(i)]^{\mathrm{T}}$ which are given as follows:

$$\mathbf{CL}(i) = [x(i)\ y(i)\ z(i)\ n_x(i)\ n_y(i)\ n_z(i)]^{\mathrm{T}} \tag{1.1}$$

$$\{n_x(i)\}^2 + \{n_y(i)\}^2 + \{n_z(i)\}^2 = 1 \tag{1.2}$$

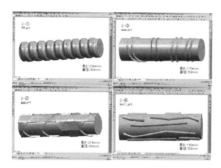

Figure 1.3. *Artistic paint roller models designed by 3D CAD*

Finally, the post-processor generates corrected NC data for the five-axis NC machine tool with a tilting head, only by considering the tool length [NAG 96]. Figure 1.4 shows the main head with a tilting mechanism, which has a ball-end mill at the tip. The tool length $L1 + L2$ is defined as the distance from the center of swing to the tip of the ball-end mill, which should be measured in advance. The post-processor transforms the CL data into NC data for the five-axis NC machine tool as shown in Figure 1.1. The main head can incline and rotate within the range ±90 and ±180 degrees, respectively. The inclined and rotated axes are called the 4th(B) axis and 5th(C) axis, respectively. The i-th step in the corrected NC data is written by:

$$\mathbf{NC}(i) = [\tilde{x}(i)\ \tilde{y}(i)\ \tilde{z}(i)\ b(i)\ c(i)]^{\mathrm{T}} \tag{1.3}$$

$$\tilde{j}(i) = j(i) + n_j(i)(L1 + L2),\ \ j = x, y, z \tag{1.4}$$

where $b(i)$ and $c(i)$ are the head angles of inclination and rotation, respectively. The CL data generated from the main-processor of CAM are composed of sequential points on the model's surface. If the NC data are transformed from the CL data without considering the tool length and are given to the NC machine tool, then the center of swing directly follows the NC data. This would cause a serious and dangerous interference between the main head and the workpiece. On the contrary, if the center of swing follows the corrected NC data given by equations [1.3] and [1.4], the tip of the ball-end mill can desirably move along the model surface.

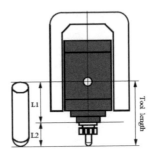

Figure 1.4. *Definition of tool length in case of a tilting head*

Recently, such a five-axis NC machine tool with a tilting head has become a center of attraction in wood product manufacturing. However, it has been recognized from machining experiments that the five-axis NC machine tool is not suitable for carving roller models as shown in Figure 1.3. Although high machining performance is expected to exceed the capability of standard NC machine tools, it has been hardly used for carving roller models yet. The main reason is that the roller models as shown in Figure 1.3 cannot be machined in a single path for the cutter, that is, models have to be machined using multiple paths for the cutter, requiring complicated cutter repositioning; thus, the 3D machining for artistic design paint rollers is complicated and time consuming. Furthermore, it is not easy to realize the modeling of a relief design on a cylindrical shape. To overcome these problems, a 3D machining system based on a three-axis NC machine tool with a rotary unit is discussed in the next section.

1.3. Three-axis NC machine tool with a rotary unit

1.3.1. *Introduction*

As described in the previous section, the wooden paint rollers have a cylindrical shape with an artistic design so that it is not easy to observe its elaborate carving even when the latest woodworking machinery such as a five-axis NC machine tool with a tilting head is used. Generally, metallic cylindrical parts are precisely processed by an expensive CNC turning center with milling capability. However, in the furniture manufacturing industry, three-axis NC machine tools and five-axis NC machine tools with a tilting head have been mainly used. Thus, a new woodworking machinery for wooden paint rollers should be designed by utilizing such existing NC machine tools and by considering the equipment cost. In this section, a three-axis NC machine tool with a rotary unit and its post-processor are proposed to efficiently produce artistic wooden paint rollers of many kinds of designs. The proposed machining system is used easily and at low cost by only adding a compact rotary unit on a conventional three-axis NC machine tool. The post-processor provides two effective functions. One is a transformation technique from CL data without feedrate values to NC data, mapping the y-directional pick feeds to rotational angles of the rotary unit. The post-processor allows a well-known three-axis NC machine tool to easily transcribe a relief design from a flat model to a cylindrical model. The other function is an elaborate addition of feedrate codes according to the curvature of each design to protect the cylindrical surface from an undesirable edge chipping. The post-processor generates safe feedrate values using a simple fuzzy reasoning method while checking edges and curvatures in a relief design, and appends them into NC data. The post-processed NC data mildly act on the fragile edges of wooden paint rollers [NAG 09].

1.3.2. *Post-processor for a three-axis NC machine tool with a rotary unit*

Conventional paint rollers generally have no artistic designs or if they do, designs are limited to flat or simple patterns even if they have. As mentioned in the previous section, unfortunately, it is not easy to carve a relief design on a cylindrical workpiece even if the five-axis NC machine tool with a tilting head is used. First of all, we discuss the problem concerning 3D machining of wooden cylindrical shapes with a relief design. When the modeling of a roller is conducted using a 3D CAD, a base cylindrical shape is modeled in advance. Then a favorite relief design is drawn on the cylindrical model. However, the modeling of relief design on the cylindrical shape as shown in Figure 1.3 is a complicated task, even when high-end 3D CAD software is used. Next, it is not easy to realize its 3D machining using the five-axis NC machine tool with a tilting head, in which the NC data generated from CAM are composed of x-, y-, z-, b-, and c-directional components. Thus, to easily provide many kinds of paint rollers with a wide variety and low volume to home making industries, a machining system that can directly carve an artistic relief design on a cylindrical workpiece must be developed.

To cope with these problems, a new 3D machining system is considered based on a three-axis NC machine tool with a rotary unit, and a post-processor is proposed for the rotary unit. The post-processor allows conventional woodworking machinery as the three-axis NC machine tool to elaborately produce wooden paint rollers. An artistic design drawn on a flat model surface can easily be transcribed to a cylindrical model surface. In the remainder of this section, the system is described in detail. A three-axis NC machine tool with x-, y-, and z-axes must be first prepared to realize the proposed concept. As an example, an NC machine tool MDX-650A provided by Roland D.G. as shown in Figure 1.5 is used for experimentation. The NC machine tool is equipped with an auto tool changer ZAT-650 and a rotary unit ZCL-650A. The mechanical resolution of the rotary unit is about 0.0027 degrees.

The NC machine tool has four DOF, that is, three translations and one rotation. This section addresses how to easily make a wooden paint roller with an artistic relief design. The most important point is that proper NC data for the NC machine tool with a rotary unit can be generated in one step. To meet this end, the post-processor generates the NC data that transcribe the design on a flat model to it on a cylindrical model. By applying the post-processed NC data, the NC machine tool can directly carve an artistic relief design on a cylindrical workpiece.

Figure 1.5. *Three-axis NC machine tool MDX-650 with a rotary unit*

Next, we describe the feature of the post-processor. A desired relief design is first modeled on a flat base model. The CL data are then generated with a zigzag path as shown in Figure 1.6. In this case, the coordinate system should be set so that the pick-feed direction is parallel to the table slide direction of the NC machine tool, that is, y-direction. The proposed post-processor transforms the CL data without "FEDRAT/" statements into the corresponding NC data, mapping y-directional positions to rotational angles of the rotary unit. As can be seen from the components of the NC data, when the rotary unit is active, the table slide motion in y-direction is inactive. The post-processor first checks all steps in the CL data, and extracts the minimum value $y_{\_min}$ and the maximum value $y_{\_max}$ in y-direction. The angle $a(i)$ for the rotary unit at the i-th step is easily calculated from equation [1.5]:

$$a(i) = \frac{360 \times \{y(i) - y_{\_min}\}}{y_{\_length}} \qquad [1.5]$$

where $y_{\_length}$ is the length in y-direction, which is obtained by $y_{\_max} - y_{\_min}$. The CL data $\mathbf{p}(i) = [x(i)\ y(i)\ z(i)]^{\mathrm{T}}$ at the i-th step is transformed into the NC data composed of $[x(i)\ a(i)\ z(i)]^{\mathrm{T}}$ using equation [1.5]. The length in y-direction is transformed into the circumference of the roller model. The amount of the small angle $a(i)$ depends on the ratio of $y_{\_length}$ to $y_{\_pick}$. It is expected that the relief design shown in Figure 1.6 is desirably carved on the surface of a cylindrical workpiece.

Thus, the proposed system provides a function that easily transcribes an artistic design from the surface on a flat model to a surface on a cylindrical wooden workpiece fixed to the rotary unit.

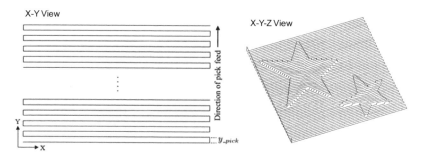

Figure 1.6. *Generation of zigzag path*

1.3.3. *Experiment*

In the previous section, a three-axis NC machine tool with a rotary unit and its post-processor were introduced to efficiently machine cylindrical wooden workpieces. The machined workpiece can be used as an artistic paint roller, which is very useful and convenient to directly transcribe a relief design to a wall just after painting. In this section, an actual machining experiment of a cylindrical wooden workpiece is conducted using the proposed system. Although up till now there have been few studies to investigate optimal machining conditions with respect to the machined curved surface of wood, Fujino *et al.* [FUJ 03] examined the influences of machining conditions such as feedrate and feed direction to the grain using two wood species, in which constant feedrate values under 2,000 mm/min were evaluated. The roughness of the machined surface was measured based on 3D profiles obtained by laser scanning, where the surface roughness was shown to be improved as the feedrate decreased. It is expected from the above result that the decrease of feedrate will be effective to suppress an undesirable edge chipping. Wooden materials are essentially brittle compared with metallic materials so that the increase of feedrate tends to bring out the edge chipping. Such a characteristic was confirmed in preliminarily conducted machining test, in which the feedrate was varied using an override function. The override function allows the NC machine tool to manually increase or decrease the programmed feedrates written with "F" code.

Figure 1.7 shows a machining scene example of a paint roller without any undesirable edge chipping, in which the feedrate values with "F" code are outputted online from the fuzzy feedrate generator [NAG 09]. The kind of wooden material used is a glued laminated wood. The maximum cutting depth of material removed is

3 mm. The wooden material was machined with the design as shown in Figure 1.6 giving the maximum cutting depth of 3 mm. An undesirable edge chipping occurred when feedrate values higher than about 800 mm/min were given. To suppress the edge chipping around the edges, the feedrates less than 800 mm/min were given by the fuzzy feedrate generator.

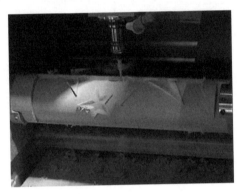

Figure 1.7. *Machining scene of a wooden paint roller*

1.4. Articulated-type industrial robot

1.4.1. *Introduction*

Industrial robots have drastically rationalized many kinds of manufacturing processes in industrial fields. The user interface provided by the robot maker has been almost limited to a teaching pendant. The teaching pendant is a useful and safe tool to obtain the position and orientation at the tip of a robot arm along a desired trajectory, but the teaching process is a very complicated and time-consuming task. In particular, when the desired trajectory includes a curved line, many through points have to be recorded in advance; the task is not easy.

For this decade, open architectural industrial robots as shown in Figure 1.8 have been produced from several industrial robot makers such as KAWASAKI Heavy Industries, Ltd, MITSUBISHI Heavy Industries, Ltd, YASKAWA Electric Corp., and so on. Open architecture, described in this section, means that the servo system and kinematics of the robot are technically opened so that various applications required in industrial fields can be planned and developed on the user's side. For example, non-taught operations by collaborating with a CAD/CAM system can be considered due to the opened accurate kinematics. Also, force control strategy using a force sensor can be implemented due to the opened servo system. In this section, a 3D robot sander and a mold-polishing robot are introduced for wooden workpieces and metallic molds, respectively.

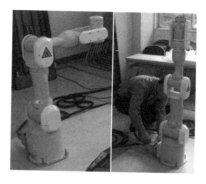

Figure 1.8. *Open architecture industrial robot Mitsubishi PA10*

1.4.2. *For sanding a wooden workpiece*

1.4.2.1. *Surface-following control*

The proposed robotic sanding system has two main features. One is that neither conventional complicated teaching tasks nor a post-processor (CL data → NC data) is required; the other is that the polishing force acting on the sanding tool and the tool's position/orientation are simultaneously controlled along a free-formed curved surface. In this section, a surface-following control method indispensable for realizing the features is described in detail. A robotic sanding task needs a desired trajectory so that the sanding tool attached to the tip of the robot arm can follow the object's surface, keeping contact with the surface from the normal direction. In executing a motion using an industrial robot, the trajectory is generally obtained in advance, for example, through a conventional robot teaching process. When the conventional teaching for an object with complex curved surface is conducted, the operator has to input a large number of teaching points along the surface. Such a teaching task is complicated and time-consuming.

Next, a sanding strategy dealing with the polishing force is described in detail. The polishing force vector $\mathbf{F}(k) = [F_x(k)\ F_y(k)\ F_z(k)]^{\mathrm{T}}$ is assumed to be the resultant force of contact force vector $\mathbf{f}(k) = [f_x(k)\ f_y(k)\ f_z(k)]^{\mathrm{T}}$ and kinetic friction force vector $\mathbf{F}_r(k) = -[F_{rx}(k)\ F_{ry}(k)\ F_{rz}(k)]^{\mathrm{T}}$ that are given to the workpiece as shown in Figure 1.9, where the sanding tool is moving along the surface from (A) to (B). $\mathbf{F}_r(k)$ is written as follows:

$$\mathbf{F}_r(k) = -\mathrm{diag}(\mu_x, \mu_y, \mu_z)\|\mathbf{f}(k)\|\frac{\mathbf{v}_t(k)}{\|\mathbf{v}_t(k)\|} - \mathrm{diag}(\eta_x, \eta_y, \eta_z)\mathbf{v}_t(k) \qquad [1.6]$$

where the first term is the Coulomb friction and the second is the viscous friction. μ_i and η_i ($i = x$, y, z) are the i-directional coefficients of Coulomb friction per unit contact force and of viscous friction, respectively. Each friction force is generated by $\mathbf{f}(k)$ and $\mathbf{v}_t(k)$, respectively. The polishing force $\mathbf{F}(k)$ is represented by:

$$\mathbf{F}(k) = \mathbf{f}(k) - \mathbf{F}_r(k) \qquad [1.7]$$

The polishing force magnitude can easily be measured using a three DOF force sensor attached between the tip of the arm and the sanding tool, which is obtained by:

$$\left\| \mathbf{F}(k) \right\| = \sqrt{\left(^S F_x(k)\right)^2 + \left(^S F_y(k)\right)^2 + \left(^S F_z(k)\right)^2} \qquad [1.8]$$

where $^S F_x(k)$, $^S F_y(k)$, and $^S F_z(k)$ are the directional components of force sensor measurements in the sensor coordinate system. The force sensor used is the NITTA IFS-67M25A with a sampling rate of 8 kHz. Although the IFS-67M25A is a six DOF force/moment sensor, the moment components are ignored because the moment data are not needed in the force control system. The error $E_f(k)$ of polishing force magnitude is calculated by:

$$E_f(k) = \left\| \mathbf{F}(k) \right\| - F_d \qquad [1.9]$$

where F_d is the desired polishing force.

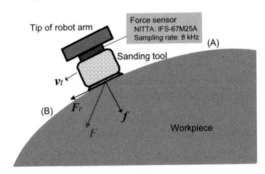

Figure 1.9. *Polishing force **F** composed of contact force **f** and friction force **F**ᵣ*

1.4.2.2. *Feedback control of polishing force*

In the wooden furniture manufacturing industry, skilled workers usually use hand-held air-driven tools to finish the surface after machining or painting. These types of tools cause large magnitude vibrations with high frequency so that it is difficult for a skilled worker to sand the workpiece keeping a desired polishing force. Consequently, undesirable unevenness tends to appear on the sanded surface. To achieve good surface finishing, it is fundamental and effective to stably control the polishing force. When the robotic sanding system runs, the polishing force is controlled by the impedance model following force control with integral action given by:

$$v_{normal}(k) = v_{normal}(k-1)e^{(-B_d/M_d)\Delta t} - \left(e^{(-B_d/M_d)\Delta t} - 1\right)\frac{K_f}{B_d}E_f(k) + K_{fi}\sum_{n=1}^{k}E_f(n) \quad [1.10]$$

where $v_{normal}(k)$ is the velocity scalar; K_f is the force feedback gain; K_{fi} is the integral control gain; and M_d and B_d are the desired mass and desired damping coefficients, respectively. Δt is the sampling width. Using $v_{normal}(k)$, the normal velocity vector $v_n(k) = [v_{nx}(k)\ v_{ny}(k)\ v_{nz}(k)]$ at the contact point is represented by:

$$v_n(k) = v_{normal}(k)\frac{o_d(k)}{\|o_d(k)\|} \quad [1.11]$$

where $o_d(k)$ is the normal vector at the contact point, which is obtained from CL data.

1.4.2.3. *Feedforward and feedback control of position*

Currently, wooden furniture is designed and machined with 3D CAD/CAM systems and NC machine tools, respectively. Accordingly, the CL data generated from the main-processor of CAM can be used for the desired trajectory of the sanding tool. The block diagram of the surface-following controller implemented in the robot sander is shown in Figure 1.10. The position and orientation of the tool attached to the tip of the robot arm are feedforwardly controlled by the tangent velocity $v_t(k)$ and rotational velocity $v_r(k)$, respectively, referring to the desired position $x_d(k)$ and desired orientation $o_d(k)$. $v_t(k)$ is given through an open-loop action so as not to interfere with the force feedback loop. The polishing force is regulated by $v_n(k)$ which is orthogonal to $v_t(k)$. $v_n(k)$ is given to the normal direction referring to the orientation vector $o_d(k)$.

It should be noted, however, that using only $v_t(k)$ is not enough to precisely carry out desired trajectory control along CL data: actual trajectory tends to deviate from the desired one so that the constant pick feed (e.g. 20 mm) cannot be performed. This undesirable phenomenon leads to a lack of evenness on the surface. To overcome this problem, a simple position feedback loop with small gains is added as shown in Figure 1.10 so that the tool does not seriously deviate from the desired pick feed. The position feedback control law generates another velocity $v_p(k)$ given by:

$$v_p(k) = S_p\left\{K_p E_p(k) + K_i\sum_{k=1}^{n}E_p(n)\right\} \quad [1.12]$$

where $\mathbf{S}_p = \mathrm{diag}(S_{px}, S_{py}, S_{pz})$ is a switch matrix to realize a weak coupling control in each direction. If $\mathbf{S}_p = \mathrm{diag}(1, 1, 1)$, then the coupling control is active in all directions; whereas if $\mathbf{S}_p = \mathrm{diag}(0, 0, 0)$, then the position feedback loop does not contribute to the force feedback loop in all directions. $\mathbf{E}_p(k) = \mathbf{x}_d(k) - \mathbf{x}(k)$ is the position error vector. $\mathbf{x}(k)$ is the current position of the sanding tool attached to the tip of the arm and is obtained from the forward kinematics of the robot. $\mathbf{K}_p = \mathrm{diag}(K_{px}, K_{py}, K_{pz})$ and $\mathbf{K}_i = \mathrm{diag}(K_{ix}, K_{iy}, K_{iz})$ are the position and integral gain matrices, respectively. Each component of $\mathbf{K}_p$ and $\mathbf{K}_i$ has to be set to small values so as not to obviously disturb the force control loop. Finally, recomposed velocities $\tilde{\mathbf{v}}_n(k) = [\mathbf{v}_n^{\mathrm{T}}(k)\ 0\ 0\ 0]^{\mathrm{T}}$, $\tilde{\mathbf{v}}_t (k) = [\mathbf{v}_t^{\mathrm{T}}(k)\ \mathbf{v}_r^{\mathrm{T}}(k)]^{\mathrm{T}}$, and $\tilde{\mathbf{v}}_n(k) = [\mathbf{v}_p^{\mathrm{T}}(k)\ 0\ 0\ 0]^{\mathrm{T}}$ are summed up to make a velocity command $\mathbf{v}(k)$, and the $\mathbf{v}(k)$ is given to the reference of the Cartesian-based servo controller of the industrial robot. It is known that the complete six constraints, which consist of three DOF positions and three DOF forces in a constraint frame, cannot be simultaneously satisfied. However, the delicate cooperation between the position feedback loop and force feedback loop is an important key point to successfully achieve robotic sanding with curved surface.

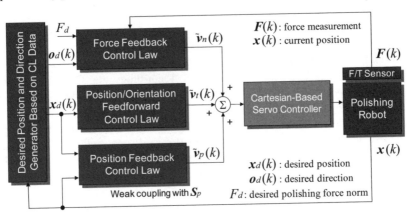

Figure 1.10. *Block diagram of surface-following controller based on CL data*

1.4.2.4. *Experiment*

In this section, experimental results of surface sanding are shown using the proposed robot sander. The overview of the robot sander developed based on KAWASAKI FS20 is shown in Figure 1.11. The orbital sanding tool is widely used by skilled workers to sand or finish a workpiece with curved surface. The base of the orbital sanding tool can perform eccentric motion. It is the reason why it is not only a powerful sanding tool but also gives good surface quality with less scratch. In this experiment, an orbital sanding tool is selected and attached to the tip of the robot arm via a force sensor. The diameter of the circular base and the eccentricity are

90 mm and 4 mm, respectively. The weight of the sanding tool is about 1.5 kg. When a sanding task is conducted, a circular pad with a sanding paper is attached to the base.

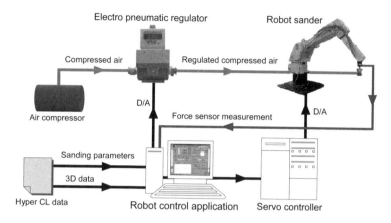

Figure 1.11. *3D robot sander with peripheral devices*

Figure 1.12 shows the sanding scene using the robot sander. In this case, the polishing force was satisfactorily controlled around a desired value. Hand-held air-driven tools are usually used by skilled workers to sand wooden material used for furniture. These types of tools cause large noise and vibration. Furthermore, the system of force control consists of an industrial robot, force sensor, attachment, hand-held air-driven tool, zig, and wooden material. Because each of them has a stiff property, it is not easy to strictly keep a desired polishing force without overshoot and oscillation. This is the reason why measured values of the polishing force tend to have spikes and noise. However, the result is much better than that by a skilled worker. Although it is difficult for a skilled worker to simultaneously keep the desired polishing force, tool position, and orientation even for a few minutes, the robot sander can perform the task more uniformly and perseveringly.

Figure 1.13a shows the target workpiece after NC machining, that is, before sanding, which is a representative shape that the conventional sanding machines cannot sand sufficiently. The pick feed in the NC machining is set to 3 mm. The surface before sanding has undesirable cusp marks higher than 3 mm for every pick feed. The robot sander first removes the cusp marks using a rough sanding paper #80, then sands the surface using a sanding paper of a middle roughness #220, and finally a smooth paper #400. The diameters of the pad and paper are cut to 65 mm, which are larger than that of the ball-end mill (17 mm) used in the NC machining process. The pad was put between the base and the sanding paper. The CL data were regenerated with a pick feed of 15 mm for the robotic sanding. The contour was made so as to be a small size with an offset of 15 mm to prevent the edge of the workpiece from over sanding. Table 1.1 shows the other sanding conditions and the parameters of the surface-following controller. These semi-optimum values were found through trial and error. Figure 1.13b shows the surface after the sanding process. The touch feelings with both the fingers and the palm were very satisfactory. Undesirable cusp marks were not observed at all. There was also no over sanding around the edge of the workpiece and no swell on the surface. Furthermore, we conducted a quantitative evaluation using a stylus instrument so that the measurements obtained by the arithmetical mean roughness (Ra) and max height (Ry) were around 1 µm and 3 µm, respectively. Figure 1.14 shows a piece of artistic furniture using the workpiece sanded by the robot sander. It was confirmed from the experimental result that the proposed robotic sanding system could successfully sand the wooden workpiece with curved surface.

Figure 1.12. *Sanding scene of a curved workpiece*

Figure 1.13. *Curved surfaces before and after sanding process*

Conditions or parameters	Values
Robot	KAWASAKI FS3OL
Force sensor	NITTA IFS-100M40A
Workpiece	Japanese oak
Size (mm)	1,200 x 425 x 85
Diameter of sand paper (mm)	65
Grain size of sand paper (#)	80→220→400
Desired polishing force $\mathbf{F}_d$ (kgf)	1.0
Feed rate $\|\mathbf{v}_t\|$ (mm/s)	30
Pick feed of CL data (mm)	15
Air pressure of orbital sanding tool (kgf/cm^2)	4.0
Desired mass coefficient $\mathbf{M}_d$ (kgf.s^2/mm)	0.01
Desired damping coefficient $\mathbf{B}_d$ (kgf.s/mm)	20
Force feedback gain $\mathbf{K}_f$	1
Integral control gain for polishing force $\mathbf{K}_{fi}$	0.001
Switch matrix for weak coupling control .$\mathbf{S}_p$	diag(0, 1, 0)
Position feedback gain matrix $\mathbf{K}_p$	diag(0, 0.01, 0)
Integral control gain matrix $\mathbf{K}_i$, for position	diag(0, 0.0001, 0)
Sampling width Δt (ms)	0.01

Table 1.1. *Sanding conditions and control parameters*

Courtesy of Workshop Nishida in Okawa City

Figure 1.14. *Artistic furniture using the workpiece sanded by the robot sander*

1.4.3. *For mold finishing*

1.4.3.1. *Introduction*

In the next stage, we try to include an industrial robot into the polishing process of polyethylene terephthalate (PET) bottle molds. As can be guessed, the sizes of the target workpieces are smaller than parts constructing furniture, that is, the radius of curvature is also smaller. In the manufacturing industry of PET bottle molds, 3D CAD/CAM systems and machining centers are similarly used, and these advanced systems have been drastically rationalizing the design and manufacturing process of metallic molds. On the contrary, most of the polishing processes after a machining process have been supported by skilled workers with capabilities concerning both dexterous force control and skillful trajectory control for an abrasive tool. The skilled workers usually use mounted abrasive tools of several sizes and shapes. In using these types of tools, keeping contact with the metallic workpiece with a desired contact force and a tangential velocity is the most important factor to obtain a high-quality surface. When performing a polishing task, it is also a key point that skilled workers reciprocatingly move the abrasive tool back and forth along the object surface.

Since the repetitive position accuracy at the tip of articulated-type industrial robots is 0.1 mm or thereabouts [NAG 08], it is very difficult to polish the surface of the metallic mold using only position control strategy. In the polishing process of PET bottle molds, the surface accuracy Ra of 0.1 μm or less is finally required for mirror-like finishing. In particular, when an industrial robot makes contact with a metallic workpiece, several factors that decrease the total stiffness of the system are included. They are called backlash, strain, and deflection, all of which exist not only in the robot itself but also in the force sensor, abrasive tool, jig, base frame, and so on. Therefore, it is meaningless to discuss the position accuracy at the tip of the abrasive tool attached to the robot arm. If position control is used for a polishing task where an abrasive tool and metallic workpiece contact each other, then both the stiffness of the robot and the total stiffness including the abrasive tool have to be extremely high. However, this problem and the problem on uncertainty of workpiece positioning have not been overcome.

It is actually known that no advanced polishing robots have been successfully developed yet on a commercial basis for such metallic molds with curved surface as in the case of PET bottle molds, due to the poor polishing quality and the complicated operation. The reasons why conventional polishing robots based on an industrial robot could not satisfactorily finish the curved surface of molds are listed as follows:

– Conventional industrial robots provide only a teaching pendant as a user-interface device. Precise teaching along a curved surface is extremely difficult and complicated.

– Kinematics and servo control, which are indispensable in developing a real-time application for mold polishing, have not been technically available as an open source to engineers and researchers.

– No successful control strategy has been proposed yet for mold polishing with curved surface. Compatibility between force control and position control is needed for higher surface quality.

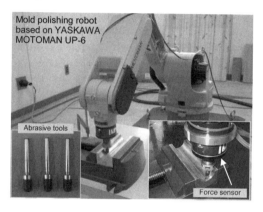

Figure 1.15. *Mold-polishing robot developed based on MOTOMAN UP6*

In this section, dexterous techniques are presented for understanding a skillful mold-polishing robot as shown in Figure 1.15. The CL data with normal vectors called multi-axis CL data can be used for not only a desired trajectory of tool translational motion but also a contact direction given to a mold. The impedance model following force control method keeps the polishing force, composed of a contact force and kinetic friction force, constant. A CAD/CAM-based position/force controller in Cartesian space by referring to such multi-axis CL data is proposed for the polishing robot with a ball-end abrasive tool. The surface polishing is achieved by controlling both the tool position along the CL data and the polishing force. The difference with the block diagram shown in Figure 1.10 is that the orientation of the tool is always fixed to z-axis in work coordinate system; the orientation component in CL data is used for the force direction to be given to a mold. The CAD/CAM-based position/force controller is applied to an industrial robot with an open architecture controller. The effectiveness and validity of the mold-polishing robot with the CAD/CAM-based position/force controller are demonstrated through an actual polishing experiment.

1.4.3.2. Basic polishing scheme for a ball-end abrasive tool

In this section, a control strategy that efficiently uses the contour of a ball-end abrasive tool is introduced for the mold polishing with curved surface. In polishing, the polishing force acting between the abrasive tool and the target mold is controlled. The polishing force is the most important physical factor that largely affects the quality of polishing, and assumed to be the resultant force of contact and kinetic friction forces. The mold-polishing robot is shown in Figure 1.15, in which a ball-end abrasive tool with a radius of 5 mm is attached to the tip of a six DOF-articulated industrial robot through a force sensor. The abrasive tool is generally attached to a portable electric sander so that the power of polishing is obtained by its high rotational motion, for example, 10,000 rpm. In this case, however, it is very difficult for a skilled worker to keep regulating suitably the power, contact force, and tangential velocity for many minutes according to an object's shape, and so, undesirable over-polishing tends to occur frequently. Thus, to protect the mold surface against the over-polishing, the proposed polishing robot keeps the tool's rotation slow, and polishes the mold using the resultant force $\mathbf{F_r}$ of the Coulomb friction and the viscous friction. Each friction force is generated by the contact force $\mathbf{f}$ in normal direction and the tangent velocity $\mathbf{v_t}$, respectively.

Figure 1.16 shows the control strategy considering the kinetic friction forces. In this figure, $\mathbf{f}$ is given by the normal velocity $\mathbf{v_n}$ at the contact point between the abrasive tool and the mold. $\mathbf{v_n}$ is yielded by the force controller given by equation [1.11]. In this chapter, the polishing force is defined as the resultant force of $\mathbf{F_r}$ and $\mathbf{f}$, which can be measured by a force sensor. Figure 1.17 shows an example of a force sensor, ATI Mini40 six-axis force/torque sensor. It is assumed that the polishing is well performed by a hybrid control of the tool position and the polishing force. To avoid the interference between the abrasive tool and the mold, the orientation of the tool is not going to change and is always fixed to the z-axis in the work coordinate system. Fortunately, since PET bottle molds have no over-hang, a suitable contact point between the ball-end abrasive tool and the mold can always be obtained. The proposed polishing robot does not need to use any complex tools, vision sensors, teaching systems, and jigs so that it can be used in a simple manner.

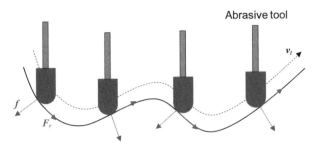

Figure 1.16. *Polishing strategy taking account of kinetic friction forces*

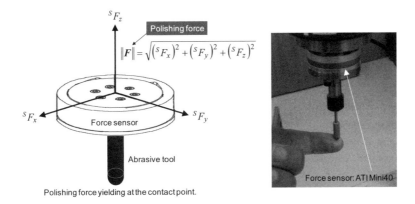

Figure 1.17. *Polishing force measured by a force sensor*

1.4.3.3. *Feedforward and feedback control of tool position*

Currently, metallic molds for PET bottle manufacturing are designed and machined with 3D CAD/CAM systems and machining centers, respectively. Accordingly, multi-axis CL data generated from the main-processor of CAM can be used for the desired trajectory of an abrasive tool. The block diagram of the CAD/CAM-based position/force controller implemented in the mold-polishing robot is almost the same as Figure 1.10, except that the tool axis is always fixed to the *z*-axis in the work coordinate system. In other words, the tool orientation is not changing to maintain the force control stability and to uniformly abrade the contour of the ball-end abrasive tool.

The position of the abrasive tool is feedforwardly controlled by the tangential velocity $\mathbf{v}_t(k)$ given by:

$$\mathbf{v}_t(k) = \mathbf{v}_{\text{tangent}} \frac{\mathbf{x}_d(k) - \mathbf{x}_d(k-1)}{\left\| \mathbf{x}_d(k) - \mathbf{x}_d(k-1) \right\|} \qquad [1.13]$$

where $\mathbf{v}_{\text{tangent}}$ is the velocity scalar which means the feedrate. $\mathbf{v}_t(k)$ is given through an open-loop action so as not to interfere with the normal velocity $\mathbf{v}_n(k)$. On the other hand, the polishing force is controlled by $\mathbf{v}_n(k)$ which is orthogonal to $\mathbf{v}_t(k)$. $\mathbf{v}_n(k)$ is given to the normal direction referring to $\mathbf{o}_d(k)$. It should be noted, however, that using only $\mathbf{v}_t(k)$ is not enough to execute desired trajectory control along the CL data, that is, the tool is not able to conduct regular pick-feed motion, for example, with a given pick feed of 0.1 mm. To avoid this undesirable phenomenon, a simple position feedback loop with a small gain is added as shown in Figure 1.10 so that the abrasive tool does not deviate from the desired pick-feed motion. The position feedback control law generates another velocity $\mathbf{v}_p(k)$ given by equation [1.12]. Finally, the velocities $\mathbf{v}_n(k)$, $\mathbf{v}_t(k)$, and $\mathbf{v}_p(k)$ are summed up, and are given to the reference of the Cartesian-based servo controller of the industrial robot. The CAD/CAM-based position/force controller neither deals with the moment nor with the rotation, and also the origin of the constraint space (force space) is always chosen at the contact point. Accordingly, although a paper by Duffy states the fallacy of modern hybrid control theory such as dimensional inconsistency, dependence on the choice of origin of the coordinates [DUF 90], our proposed system is not affected.

1.4.3.4. *Experiment*

To evaluate the validity and effectiveness of the mold-polishing robot using the CAD/CAM-based position/force controller, a fundamental polishing experiment is conducted using an aluminum mold machined by a machining center. The objective of the fundamental polishing is to remove all cusp marks on the curved surface whose heights are around 0.3 mm. The fundamental polishing before finishing process is one of the most important processes to create the best appearance for mirror-like surfaces. If the undesirable cusp marks are not uniformly removed in advance, then it is very difficult to finish the mold with a mirror-like surface without scratches, swells, and over-polishing; however, much time would be spent for the finishing process.

Figure 1.18 shows the mold-polishing robot developed based on an industrial robot KAWASAKI FS03 with open control architecture. The industrial robot provides several useful Windows API functions such as a Cartesian-based servo control and forward/inverse kinematics. A ball-end abrasive tool is attached to the tip of the robot arm via a force sensor. The surface was polished through three steps, making the grain size of the abrasive tool gradually smaller, that is, from #220, #320 to #400. Figure 1.19 shows the polishing scene using the proposed robot. When the polishing robot runs, the abrasive tool reciprocatingly rotates with ±40 deg/sec using the sixth axis of the robot so that the tool contour can be abraded uniformly. If the abrasive tool is uniformly abraded keeping the ball-end shape, the robot can keep up the initial polishing performance. Although the tool length gradually becomes shorter due to tool abrasion, the force controller absorbs the uncertainty concerning the tool length. The y-directional position feedback loop delicately contributes to the force feedback loop to keep the constant pick feed even around the inclination part of the mold.

Figure 1.18. *Mold-polishing robot developed based on KAWASAKI FS03*

Figure 1.19. *Polishing scene of PET bottle mold*

1.5. Desktop Cartesian-type robot

1.5.1. *Background*

The finishing of an LED lens mold after a machining process requires high accuracy, delicateness, and skill such that it has not been successfully automated yet. Generally, a target LED lens mold has many concave areas precisely machined with a tolerance of ±0.01 mm as shown in Figure 1.20, where each diameter is 3.6 mm. This means that the target mold is not axis-symmetric so that conventional effective polishing systems, which can deal with only axis-symmetric workpieces, cannot be applied. Accordingly, such an axis-asymmetric lens mold is polished by a skilled worker in most cases. Skilled workers generally finish small lens molds using a wood-stick tool with diamond paste while checking the finished area through a microscope. However, the smaller the workpiece is, the more difficult the task is. In particular, it is required for an LED lens mold to handle the surface uniformly and softly so that high resolutions for both position and force are indispensable for the corresponding mechatronics system.

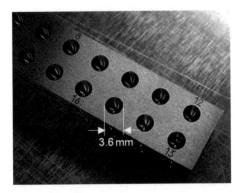

Figure 1.20. *LED lens mold*

In this section, a new desktop Cartesian-type robot with compliance controllability is presented for finishing metallic molds with a small curved surface [NAG 10]. The Cartesian-type robot consists of three single-axis devices. A wood-stick tool attached to the tip of the z-axis has a ball-end shape. Also, the control system of the robot is composed of a force feedback loop, position feedback loop, and position feedforward loop. The force feedback loop controls the polishing force consisting of tool contact force and kinetic friction force. The position feedback loop controls the position in a spiral direction. Furthermore, the position feedforward loop leads the tool tip along a spiral path. The Cartesian-type robot delicately smoothes the surface roughness at about 50 μm height on each concave area, and finishes the surface with a high quality.

1.5.2. *Cartesian-type robot*

To finish small concave areas as shown in Figure 1.20, a novel robot is proposed in this section. Figure 1.21 shows the proposed desktop Cartesian-type robot consisting of three single-axis devices with a position resolution of 1 μm. The three single-axis devices are used for x-, y-, and z-directional motions. A servo spindle motor is also used for the rotational motion of the tool axis. The servo spindle motor with a reduction gear and the tool axis work together with a belt. The size of the robot is 850 mm width, 645 mm depth, and 700 mm height. The single-axis device is a position control device ISPA with high-precision resolution provided by IAI Corp., which is composed of a base, linear guide, ball-screw, AC servo motor, and so on. The effective strokes in x-, y-, and z-directions are 400, 300, and 100 mm, respectively. The effective stiffness in the z-direction is about 177.7 N/mm when a wood-stick tool is used. Therefore, it is expected that the force resolution about 0.178 N can be performed due to the position resolution of 1 μm.

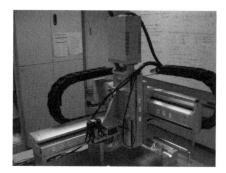

Figure 1.21. *Cartesian-type robot with compliant motion capability*

1.5.3. *Design of weak coupling control between force feedback loop and position feedback loop*

A skillful control strategy is proposed for an LED lens mold with aspherical surface. An abrasive tool moves along a spiral path as shown in Figure 1.22 while stably keeping the polishing force at a desired value. The weak coupling controller has been already proposed for the PET bottle polishing robot, in which the force feedback loop and position feedback loop are slightly coupled in a selected direction such as a pick-feed direction. In this case, a zigzag path is generally used for the desired trajectory so that a position feedback control is active only in the pick-feed direction, for example, the y-direction. On the contrary, in the case that an LED lens mold is finished by the Cartesian-type robot, the direction of the force feedback loop, that is, normal direction, changes gradually all the time when the abrasive tool contacts with the bottom center of a concave area and rises along a spiral path. Since the contact force is given to the normal direction at the contact point, the following three conditions are considered:

1. Around the bottom center of an LED lens mold, the x- and y-directional components of the normal vector are almost 0. Therefore, the force feedback control system should be constructed in the z-direction, and the position feedback control system also should be given to the x- and y-directions.

2. The z-directional component of the normal vector is almost 0 around the upper area so that the direction of the force control system periodically changes in the x- and y-directions according to the spiral path. Hence, the position feedback control system is assigned only in the z-direction.

3. Around the middle area, the x-, y-, and z-directional components of the normal vector change instantaneously along the spiral path so that the force feedback control system should be designed in the x-, y-, and z-directions. The weak coupling

control is not needed in the cases of 1 and 2. In the case of 3, however, the weak coupling control is applied to realize a regular pick-feed motion in the z-direction, while performing the stable polishing force. It is further important to independently regulate the weight of the coupling control coping with the shape of the workpiece. To deal with the problem, each component of the position feedback gain $\mathbf{K}_p = \mathrm{diag}(K_{px}, K_{py}, K_{pz})$ is varied as:

$$K_{px} = \alpha n_z \qquad\qquad\qquad [1.14]$$

$$K_{py} = \alpha n_z \qquad\qquad\qquad [1.15]$$

$$K_{pz} = \alpha(1 - n_z) \qquad\qquad\qquad [1.16]$$

where α is the basic gain for the weak coupling control. n_z $(0 \leq n_z \leq 1)$ is the z-directional component of the normal direction vector. When an abrasive tool rises along the spiral path as shown in Figure 1.22, n_z varies from 1 to 0. For example, if α is given 0.001, K_{px} and K_{py} varies from 0.001 to 0, and K_{pz} varies from 0 to 0.001 with the rise of a thin wood-stick tool.

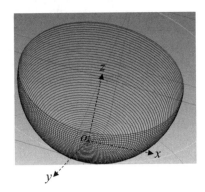

Figure 1.22. *Spiral path used for desired trajectory of LED lens mold*

1.5.4. *Frequency characteristic of force control system*

Next, the frequency characteristics of the force control system are evaluated through a simple force control experiment. A thin wood-stick tool ($\phi = 1$ mm) is used. Figure 1.23 shows an example of the desired force whose frequency is set to 1 Hz. The peak-to-peak value of desired force is 4 N. The frequency characteristics are measured within the range from 0 to 15 Hz in this order. The frequency of 0 Hz means that the desired force is set to the constant value 5 N. Figure 1.24 shows the frequency characteristics of the amplitude in the case that the wood-stick tool is used. Although 0 dB is the ideal result in the force control system, that is, the response can almost completely follow the reference shown in Figure 1.23, values over 0 dB tend to occur with the increase of the frequency. This phenomenon is caused by undesirable overshoots and oscillations in the stiff force control system. However, the desired polishing force in actual finishing tasks is generally set to a constant value, for example, 20 N so that the frequency characteristic is almost no problem.

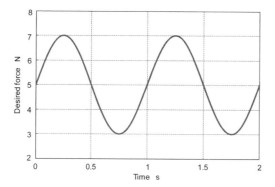

Figure 1.23. *An example of desired force*

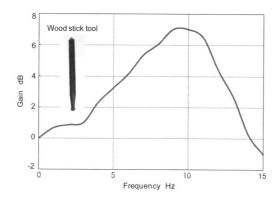

Figure 1.24. *Frequency characteristics of force control system*

1.5.5. *Finishing experiment of an LED lens mold*

In this section, the proposed system is applied to the finishing of a test workpiece of an LED lens mold. A thin ball-end tool lathed from a wooden stick is used, whose tip diameter is 1 mm. Figure 1.25 shows the finishing scene of the workpiece, where a special oil including the diamond lapping paste is being poured. Figure 1.26 shows the large-scale photos of the surfaces before and after the finishing process. It is observed that small cusps on the concave surface can be removed uniformly. The effectiveness and promise are confirmed from the finishing experiment.

Figure 1.25. *Finishing scene of a test workpiece*

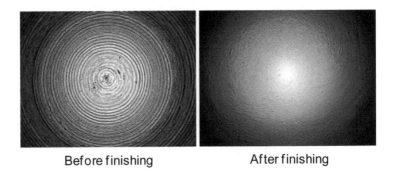

Before finishing   After finishing

Figure 1.26. *Large-scale photos of surfaces before and after finishing process*

1.6. Conclusions

Currently, 3D CAD/CAM systems are widely used in various manufacturing fields and applications. The main-processor of CAM generates cutter location data called CL data, which consists of precise position and its normal direction components. Machine tools and robots can display more superior abilities and functions by minutely using the CL data. This chapter has experimentally described and evaluated several mechatronics systems based on a 3D CAD/CAM system. A five-axis NC machine tool with a tilting head, a three-axis NC machine tool with a rotary unit, articulated-type industrial robots for furniture sanding and mold polishing, and a desktop Cartesian-type robot for LED lens mold finishing have been introduced.

Finally, we hope that the chapter will be useful not only for engineers in related industrial fields, but also for students pursuing education and research concerning NC machine tools, 3D machining, advanced control system, CAD/CAM, industrial robot, automation, and so on.

1.7. Bibliography

[DUF 90] DUFFY J., "The fallacy of modern hybrid control theory that is based on orthogonal complements of twist and wrench spaces", *Journal of Robotic Systems*, vol. 7, no. 2, 1990, pp. 139–144.

[FUJ 03] FUJINO K., SAWADA Y., FUJII Y., OKUMURA S., "Machining of curved surface of wood by ball end mill—effect of rake angle and feed speed on machined surface", *Proceedings of the 16th International Wood Machining Seminar*, part 2, 2003, pp. 532–538.

[NAG 08] NAGATA F., HASE T., HAGA Z., OMOTO M., WATANABE K., "Intelligent desktop NC machine tool with compliance control capability", *Proceedings of the 13th International Symposium on Artificial Life and Robotics*, 2008, pp. 779–782.

[NAG 09] NAGATA F., KUSUMOTO Y., WATANABE K., "Intelligent machining system for the artistic design of wooden paint rollers", *Robotics and Computer-Integrated Manufacturing*, vol. 25, no. 3, 2009, pp. 680–688.

[NAG 10] NAGATA F., MIZOBUCHI T., TANI S., HASE T., HAGA Z., WATANABE K., HABIB M.K., KIGUCHI K., "Desktop orthogonal-type robot with abilities of compliant motion and stick-slip motion for lapping of LED lens molds", *Proceedings of 2010 IEEE International Conference on Robotics and Automation (ICRA 2010)*, 2010, pp. 2095–2100.

[NAG 96] NAGATA F., WATANABE K., "Development of a post-processor module of 5-axis control NC machine tool with tilting-head for woody furniture", *Journal of the Japan Society for Precision Engineering*, vol. 62, no. 8, 1996, pp. 1203–1207 (in Japanese).

Chapter 2

Modeling and Control of Ionic Polymer–Metal Composite Actuators for Mechatronics Applications

This chapter presents the full design process through to the implementation of two innovative mechatronic devices: a stepper motor and a robotic rotary joint both with integrated soft IPMC actuators. Firstly, electromechanical modeling of the IPMC actuation response is presented. This model is then used as a tool for the mechanical design of the devices. Novel implementation of control systems to adaptively handle the highly nonlinear and time-varying response of the IPMCs and achieve successful device performance is undertaken. Experimental results are presented to validate the designs for the systems. This work demonstrates the capabilities of IPMCs and the benefits of implementing them as valid alternatives to traditional actuators.

2.1. Introduction

Ionic polymer–metal composites (IPMCs) are a novel type of smart material transducer. They are a class of electroactive polymer (EAP), acting as an actuator under the influence of an electric field and conversely producing an electric potential when mechanically deformed. Typically IPMCs have been operated in a cantilever configuration (see Figure 2.1) where a voltage is either applied or measured at the base through a set of clamped electrodes. A beam type actuation greater than 90° can be achieved with small applied voltages, typically less than 5 V. Sensing voltage

Chapter written by Andrew McDaid, Kean AW and Sheng Q. Xie.

is usually orders of magnitude lower than the voltage for actuation. This chapter focuses on IPMCs operating in actuation mode and their implementation into mechatronics systems.

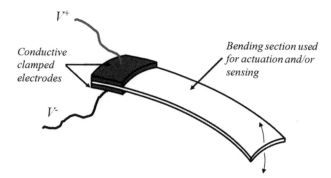

Figure 2.1. *IPMC transducer in cantilever configuration*

IPMCs are fabricated with a perfluorinated ionic membrane, for example Nafion® by DuPont, which is sandwiched between two thinly coated conducting electrodes of a noble metal, typically platinum or gold, on either side of the polymer (see Figure 2.2). The ionic polymer must be an ion exchange membrane that is permeable to cations but not anions, consisting of a fixed network of anions with mobile cations.

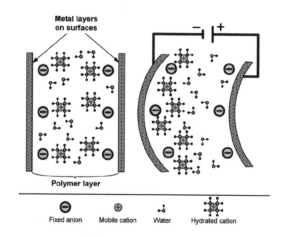

Figure 2.2. *Cross-section of IPMC transducer [AHN 10]*

By applying an electric field to the clamped electrodes a voltage is induced across the entire length of the polymer, this attracts the hydrated cations to migrate to the cathode. The accumulation of water on one side of the polymer causes expansion, which in turn produces a stress toward the anode; the uneven stress then produces a mechanical strain, or bending deformation. The opposite effect results in the sensing phenomena in IPMC. The transduction mechanism is summarized in Figure 2.3. In actuation, if a constant electric field is maintained, the actuator will eventually relax back toward the origin as the loose water diffuses back, which is known as "back-relaxation". Due to the need for hydrated ions, the IPMC operates best in an aqueous environment.

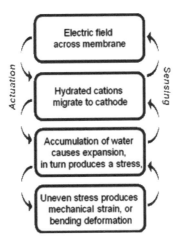

Figure 2.3. *Summary of the transduction mechanism for IPMCs*

IPMCs are most commonly manufactured in sheets and then individual transducers are cut from the sheet and as such the geometries are infinite and can be tailored to any application requirements. This makes IPMCs highly suited for implementation into any number of mechatronic systems as well as miniaturization in microelectromechanical systems (MEMs) type devices. The IPMC performance characteristics are highly dependent on their geometry; this highlights the importance of developing an accurate scalable model, as presented in the following section, for designing applications with IPMCs.

IPMC materials have a number of superior properties, in comparison with traditional actuators and other EAPs, which make them desirable for use as mechanical actuators, including:

– lightweight and thin, typically between 200 μm and 2 mm thick;

– flexible and compliant, hence safe for operating in sensitive environments;

– low power consumption, hence good for embedded and remote applications;

– achieve both micro- and macro-deflections without any gearing mechanisms [MCD 10c];

– integration of sensing and actuating using the same device;

– low actuation voltages, typically between ±1 and ±5 V;

– biocompatible and implantable in humans [SHA 01a];

– fully operational underwater, at low temperatures and in vacuum [BAR 00b, YE 08];

– completely noiseless actuation, unlike electric motors or pneumatics.

Despite these advantages there are still a number of major issues that need to be overcome before IPMCs can be widely regarded as viable alternatives for traditional actuators.

Some of the major issues include back relaxation under DC actuation, hysteresis, dehydration in air, electrolysis, non-uniform bending, extreme environmental sensitivity (hydration, temperature, humidity level [LAV 05]), and loss of mechanical force at larger displacements. All of these issues imply that the IPMCs are extremely nonlinear (especially at high inputs and low frequencies [SHA 01b, BON 07, MCD 10c]) and time varying.

Over a period of operation the highly time-varying nature of the IPMC will cause the response to change unpredictably. This cannot be fully modeled as the variance is due to ion redistribution which is a stochastic process. This random behavior also makes IPMCs very difficult to accurately control. Robust adaptive control methods must be used to make these systems reliable when operating over a period of time.

Despite the extremely complex nature of IPMCs, modeling is undertaken to give a relatively accurate representation of their response in order to aid in the design of systems and simulate their performance before implementation into real applications. Models alone, however, are not accurate enough over time to be used to develop controllers in simulation.

Some industrial applications which have been explored with IPMCs so far are a micropump [SAN 10], microgripper [YUN 06a], manipulators [HUN 08], vibration reduction [SAG 92], mobile micro-robot [TAK 06], window cleaner [BAR 00a] as well as applications in other areas like robotic finger prosthesis [CHE 09], an assistive heart compression device [SHA 01a], and a snake robot [HUN 08]. While there have been many attempts at a number of applications there are still many issues which need to be overcome before any of these devices are reliable enough for commercial deployment. Most IPMC research has been confined to laboratory experiments, while the authors of this chapter aim to solve some of the major issues to bring IPMCs into real world applications, as will be presented here.

This research deals with the implementation of IPMCs as mechanical actuators producing useful outputs in the form of displacements and forces in real world engineering systems. Implemention of the IPMC actuators into mechatronic applications with advanced adaptive control systems makes the systems "smart" as they can effectively adapt to their environment and achieve superior performance than to other comparable devices.

2.2. Electromechanical IPMC model

An accurate model describing the behavior of IPMC actuators is an essential tool for engineers designing systems incorporating IPMCs, to allow simulation and evaluation of their performance in the real world. The model should be able to predict the dynamic displacement and force output, and the relationship between them, as well as the current and hence power drawn. Also as IPMCs can be tailored to any geometry, it is very important to have a model which is geometrically scalable; this enables the appropriate size of IPMC required for the specific application to be found through simulation, before actually fabricating the actuator.

There have been a number of different IPMC models proposed in the literature, and based on their architectures they can generally be categorized into three types; white box models which attempt to model the underlying physical and chemical mechanisms of actuation, black box models based solely on system identification, or gray box models which take well-known physical phenomena of the polymer and represent them as a simple lumped parameter model [KAN 96]. Previous models have all had a number of deficiencies, white box models, for example [SHA 99, TAD 00], are typically too complex for use in practical applications, black box models are not scalable or transferable to any other IPMC [MAL 01, BON 07]. Gray box models incorporate the best capabilities and performance for aiding in mechanical design of systems. A gray box design has therefore been implemented to incorporate sufficient physical information about the polymer operating mechanisms to ensure accuracy over a number of different operating conditions and inputs, it is concise and sufficiently uncomplicated to remain practical for engineering design. The model presented here was first proposed in [MCD 09] and refined in [MCD 10b]. An overview of the key features is presented here.

The model is designed for large inputs, hence large displacements and to account for the nonlinearities at low frequencies, enabling it to be used in robotic and biomimetic applications. It has been shown in [KOT 08] that the nonlinearities are largely dependent on the level of input voltage signal, therefore a number of parameters vary with respect to the input voltage, as will be presented later.

The actuation of the IPMC has been modeled in three stages, mimicking the real physical mechanisms which cause the actuation (see Figure 2.4). A lumped parameter nonlinear electric circuit is used to predict the current absorbed by the polymer and resulting ion flux through the polymer, which is the major mechanism for actuation [SHA 01b, BON 07]. The current flow through the polymer is coupled to the ion/water flux through the electromechanical coupling term, a linear transfer function, to predict the stress induced along the polymer, σ_x, as a function of length. The stress in the polymer and any externally applied force or loads are input to the mechanical beam model which predicts the exact elastic curve of the IPMC and hence the resulting mechanical outputs, θ_x and τ_x.

Figure 2.4. *Schematic diagram of electromechanical IPMC model [MCD 10b]*

The IPMC is modeled in cantilever configuration, with one end clamped in copper electrodes (see Figure 2.5(a)). The IPMC model is split geometrically in two parts, as shown below in Figure 2.5(b), to represent the section "clamped" by the electrode and the free "beam" section. Using the electric circuit model the average current flow can be predicted for the two sections, I_C and I_B. The angular displacement, θ_T, and the blocked torque, τ_T, are the mechanical outputs.

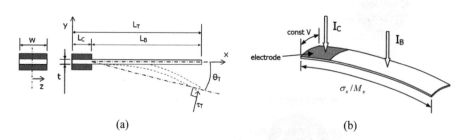

(a)   (b)

Figure 2.5. *(a) Cross-section and (b) 3D schematic of IPMC and geometric parameters [MCD 10b]*

In order for the model to be completely scalable for different dimensions of the actuator, all parameters throughout the model are expressed in terms of the IPMC geometric quantities. A list of these parameters is given in Table 2.1. All other model parameters will be introduced and their physical representation explained when they are defined in the following text.

L_T	Total length of IPMC
L_C	Length of IPMC clamped in electrodes
L_B	Length of the free "beam" section
w	Width of IPMC
t	Thickness of IPMC
θ_T	Tip angle
τ_T	Tip torque

Table 2.1. *Geometric model parameters*

2.2.1. *Nonlinear electric circuit*

It has been widely reported that the main mechanism for mechanical actuation of an IPMC is the ion and hence water flux through the polymer. It is therefore important to model the current flow in the IPMC as this can be coupled to the ion flow. Also the ability to accurately predict power consumption will be useful for designers.

Due to the characteristics of the material, resistance, capacitance, etc., the current draw is modeled using an equivalent electric circuit. The electrical response is characterized by a dynamic and steady-state response. The steady-state response is a nonlinear function of the input voltage. The dynamic response can be accurately characterized by two resistor–capacitor (RC) networks. It is evident that this capacitive dynamic response gives rise to the back relaxation phenomena as well as hysteresis in the polymer, so the circuit is capable of accurately modeling all these behaviors.

The proposed nonlinear electric circuit is shown in Figure 2.6. Researchers have proposed models using a similar approach previously [NEW 02, BON 07], also models including a nonlinear capacitance of the IPMC have been presented in [POR 08, CHE 09], but this circuit presents a number of advances over these existing models.

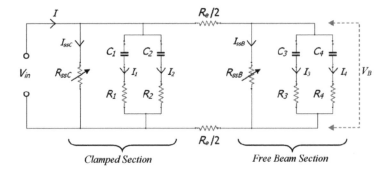

Figure 2.6. *Nonlinear electric circuit model [MCD 10b]*

Variable resistors, Rss_C and Rss_B are used to model the nonlinear steady-state absorbed current in the clamped and beam section respectively. The RC branches 1 and 2 represent the dynamic response of the polymer through the clamped section, while branches 3 and 4 represent the dynamic response through the free section of the IPMC. The two shunt resistors $R_e/2$ represent the electrode surface resistance and are the average of the ohmic resistance of the surface of the IPMC electrodes. V_B is the average voltage through the polymer thickness in the free beam section, that is after half the ohmic loss along the electrodes. The electrode resistance R_e has been measured experimentally using a four-point probe technique, and then the R_S or sheet resistance value can be calculated in "ohm/square". R_e can then be expressed in terms of the geometry of the IPMC only and hence can be scaled for different sized actuators using equation [2.1].

$$R_e = R_S \frac{L_B}{w} \qquad\qquad [2.1]$$

Rss_C and Rss_B account for the nonlinear phenomenon which occurs in the IPMC at very low frequencies and steady-state. They also incorporate the equivalent "through-resistance" of the hydrated polymer membrane. These two material properties cannot be directly experimentally measured and so are consequently combined and expressed as an equivalent resistivity, ρ_{SS}. The resistivity, which is dependent on the input voltage, is found empirically through the steady-state relationship between absorbed current and input voltage to be approximated as the third-order polynomial in equation [2.2], whose independent variable is input voltage.

$$I_{ss} = aV_{ss}^3 \qquad\qquad [2.2]$$

The values for Rss_C and Rss_B can be calculated using circuit analysis at steady-state. The resistances are then converted to an equivalent resistivity, ρ_{SS}, using equations [2.3] and [2.4] and then can be scaled to any IPMC dimensions.

$$R_{ssC} = \rho_{ss} \frac{t}{L_C w} \tag{2.3}$$

$$R_{ssB} = \rho_{ss} \frac{t}{L_B w} \tag{2.4}$$

R_1, R_2, R_3, and R_4 represent the resistance against charges flowing through the IPMC that are involved in the dynamic response. They are expressed in terms of the IPMC geometry and an equivalent resistivity in order to enable them to be scaled for different sized actuators. The portion of IPMC in the clamped and free sections have the same material properties and therefore are represented with matching resistivities ρ_f and ρ_s as in equations [2.5]–[2.8], where ρ_f represents the resistance against fast flowing charges and ρ_s represents the resistance against the slow flowing charges through the polymer material.

$$R_1 = \rho_f \frac{t}{L_C w} \tag{2.5}$$

$$R_2 = \rho_s \frac{t}{L_C w} \tag{2.6}$$

$$R_3 = \rho_f \frac{t}{L_B w} \tag{2.7}$$

$$R_4 = \rho_s \frac{t}{L_B w} \tag{2.8}$$

Similarly, capacitors C_1, C_2, C_3, and C_4 govern the time constants for the charges flowing through the IPMC that are involved in the dynamic response. They are expressed in terms of the IPMC geometry and an equivalent permittivity in order to enable them to be scaled. The portions of IPMC in the clamped and free sections have the same material properties and therefore are represented with matching permittivities ε_f and ε_s as in equations [2.9]–[2.12], where ε_f controls the time constant for the fast flowing charges and ε_s controls the time constant for the slow flowing charges through the polymer material.

$$C_1 = \varepsilon_f \frac{L_C w}{t} \tag{2.9}$$

$$C_2 = \varepsilon_s \frac{L_C w}{t} \tag{2.10}$$

$$C_3 = \varepsilon_f \frac{L_B w}{t} \tag{2.11}$$

$$C_4 = \varepsilon_s \frac{L_B w}{t} \tag{2.12}$$

Now that all the parameters for the electric circuit have been defined, the circuit can be analyzed and the current absorbed by the IPMC predicted.

2.2.2. *Electromechanical coupling*

It is widely accepted that the conversion of electrical energy to mechanical energy is due to the inner charge/water molecule redistribution [SHA 01b, BUF 08]. A number of microscopic actuation mechanisms give rise to the macroscopic deformations of the IPMC [BUF 08]. This is an extremely complex and also stochastic process, which is still not fully understood, so will introduce far too much complexity to the model. To ensure the model is practical yet still realistic, an assumption is made that the electric current flow at any point along the length of the beam can be linearly coupled to the ion/water flow and hence to a longitudinal induced stress of the IPMC beam at that point. This is physically interpreted as the amount of mass of water that flows through the thickness of the beam is directly proportional to the amount of swelling and stress in one side of the IPMC.

A number of different forms for the linear electromechanical coupling transfer function $C_{EM}(s)$ were considered and tested. Based on these tests and work in [BON 07] by Bonomo *et al.*, the most accurate response was achieved with the form shown below, equation [2.13], which includes one zero and two poles.

$$C_{EM}(s) = K \frac{s + Z}{s^2 + P1 \cdot s + P2} \tag{2.13}$$

The values K, Z, $P1$, and $P2$ are found empirically. The stress as a function of length along the IPMC can then be calculated by:

$$\sigma_x(s) = C_{EM}(s) \cdot I(x) \tag{2.14}$$

2.2.3. Mechanical beam model

The stress generated along the length of the IPMC, as a result of a voltage input, can be converted to a bending moment or electrically induced moment (EIM), using the "flexure formula", $\sigma = \dfrac{My}{I}$, where y is the distance from the neutral, x, axis in the y direction, and I is the moment of inertia about the neutral axis. It has been reported in literature that the electromechanical conversions occur at the interface between the electrode and the polymer membrane [NEM 00, NEW 02], using this fact, y is taken as $t/2$.

Now the EIM can be calculated as a function of length in the Laplace domain $\mathrm{EIM}_x(s) = \dfrac{2\sigma_x(s)I}{t}$.

The EIM and all other moments which are induced in the IPMC beam as a result of externally applied forces or loads are added to calculate the resultant total bending moment. In this way the proposed model can accommodate any external force/moment or load that acts anywhere along the length of the IPMC. This makes the model extremely useful in mechanical design.

In order to relate bending moment to the beam deflection commonly the approximation $M/EI = d^2v/dx^2$ is used by assuming a shallow curve where EI is the product of the modulus of elasticity and moment of inertia of the IPMC. When modeling the IPMC with large inputs, this assumption will not hold true, and therefore will not give an accurate representation of the true displacement of the beam. Also using this method, as has been done in previous models [NEW 02, BON 07, CHE 08], will only give a linear displacement as a function of the length and not the actual elastic curve and bending displacement. This IPMC model needs to be accurate for large displacements. Therefore to overcome this issue, the beam has been "segmented" into smaller pieces along its length. Providing the segments are small enough, they can be analyzed individually and the shallow curve assumption will hold true. Each segment will have its own elastic curve and when put together will make up the entire elastic curve of the IPMC beam. Using this method will then allow the true shape of the bending actuator to be found, an angular displacement and not a simple linear approximation as in previous models [NEW 02, BON 07, CHE 08]. An example of a segmented curve is shown in Figure 2.7(a), with a segment length of 1 mm and the combined resulting curve for the IPMC with 30 mm free length is shown in Figure 2.7(b). The individual segment displacements are more than an order of magnitude smaller than the length of segment, which ensures that the shallow curve assumption holds true.

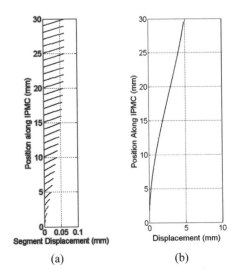

Figure 2.7. *Typical deflections for an IPMC with 30 mm free section and segment length of 1 mm (a) individual segments, (b) combined elastic curve for IPMC [MCD 10b]*

2.2.4. Parameter identification and results

All the model parameters were identified using a Nafion®-based IPMC actuator 24 mm long, 10 mm wide, and an average thickness of 0.7 mm, with Pt-coated electrodes (see Table 2.2), using solely the blocked force results. The agreement between the simulated and experimental free displacement and force at varying displacements then verifies that the model is accurate for the full actuation response of the IPMC.

R_S	$21.12 \ \Omega$
a	8.84×10^{-4}
ρ_S	$-4.3983\|Vin\| + 15.446$
ρ_f	$-1.2561\|Vin\| + 4.4083$
ε_S	$0.4420\|Vin\| + 0.3993$
ε_f	$0.1981\|Vin\| + 0.5078$
K	$-9,434.0\|Vin\| + 74,869$
Z	$0.1431\|Vin\| + 4.3595$
$P1$	$2.8180\|Vin\| + 7.7198$
$P2$	$-0.0321\|Vin\| + 0.4499$
E	$0.1757 \ GPa$
I	$0.2858 \times 10^{-12} \ m^4$

Table 2.2. *Identified model parameters*

The model with the parameters identified as in Table 2.2 is simulated for 60 seconds and both absorbed current and blocked force are compared with actual measured values to evaluate their correlation. Figures 2.8(a)–(c) plot the simulated and actual experimental results for the current draw for the actuator at 1 V, 2 V, and 3 V respectively. The plots clearly show the excellent correlation between the simulated current draw and the actual measured current draw. It can be seen that the peak current draw, the dynamic decay, and the steady-state values can all be accurately predicted by the model.

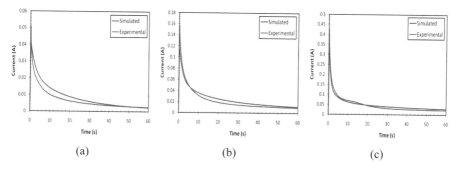

(a)   (b)   (c)

Figure 2.8. *Experimental and simulated current for (a) 1 V, (b) 2 V and (c) 3 V step inputs [MCD 10b]*

Figure 2.9(a) plots the corresponding simulated and actual measured blocked force for 1–3 V inputs at zero displacement. Again it can be seen that there is a good agreement between the simulated and experimental results. The 2 V experiment is lower than the simulated, but with the unrepeatable nature of the IPMC, some deviations can be expected and are acceptable. The results of the free deflection experiments and corresponding model simulation are shown, see Figure 2.9(b) for 1–3 V. The results show a good match to the actual measured displacements.

It can be seen that the model captures both the nonlinear steady-state characteristics, after the system has been left for a long time to settle, as well as the fast dynamic response of the IPMC at a large range of voltage inputs (up to 3 V). The model also correctly predicts the back relaxation phenomena to a DC input.

The equivalent electric circuit has also been designed to accurately account for the hysteresis effects exhibited by IPMCs [ZHE 05, PUN 07, SHA 07]. The model is simulated with a 3 V amplitude sinusoid wave of 0.2, 0.1, and 0.025 Hz, over 60 seconds. The results are shown in Figure 2.10. It can be seen that there is an obvious hysteresis loop as the current draw and tip displacement follow a different path when the voltage is increasing to when the voltage is decreasing. The hysteresis loop predicted by the model is caused by the transient behavior of the variable RC branches in the electrical model. The level of hysteresis is therefore dependent on the input voltage and frequency. This is shown in Figure 2.10 by the different paths followed for the different frequencies simulated, which is also observed in the real system. It is also clear that the simulated data indeed captures the nonlinearity.

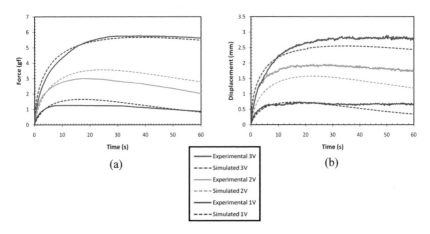

Figure 2.9. *Experimental and simulated (a) blocked force, (b) displacement for 1, 2, and 3 V step inputs [MCD 10b]*

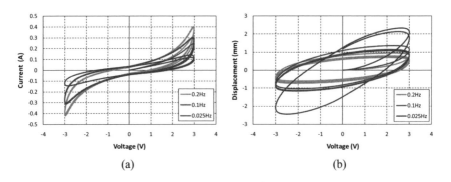

Figure 2.10. *Simulated (a) current draw and (b) tip displacement vs. input voltage for a ±3 V sinusoid wave input of 0.2, 0.1, and 0.025 Hz over 60 seconds [MCD 10b]*

The model is also capable of predicting the relationship between the force and displacement. The blocked force has been measured at a number of different displacements. Figure 2.11 shows the experimental passive blocked force (0 V input) at varying displacements and the peak blocked force for –3 V to +3 V inputs at each displacement. The model is then simulated for the same conditions. The results plotted show the close agreement between the model and the actual measurements. This demonstrates the model's ability to predict the force as a function of displacement as well as the free displacement and velocities of the IPMC, showing that the model is indeed accurate for the complete actuation response of the IPMC.

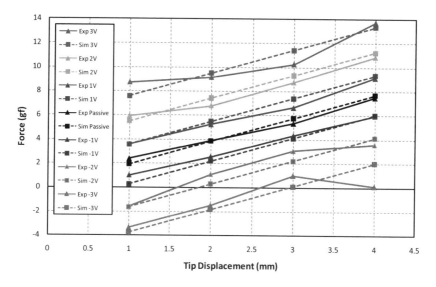

Figure 2.11. *Experimental and simulated peak blocked force at varying tip displacements for –3 V to +3 V step inputs [MCD 10b]*

All parameters have been developed with a physical interpretation and are expressed in terms of IPMC geometry to ensure scalability. Experiments were carried out using two different samples based on the same Nafion®-Pt material, one $30 \times 10 \times 0.7$ mm and another $35 \times 10 \times 0.7$ mm actuator, both with a 5 mm clamped section. Figure 2.12 compares the results of the experimental data and the model predictions for the 30 mm and 35 mm long actuators.

These plots clearly show that the complete actuation response, force, and displacement, in both the dynamic and steady-state range, of an IPMC actuator can be determined for different actuator geometries. Although there are some small deviations, it can be seen here that both the dynamic and steady-state response can be reasonably accurately simulated for force and displacement, with 1, 2, and 3 Vs.

A complete model for the actuation response has been developed which is a useful tool used to design mechatronics systems for a real life applications. In the remainder of this chapter examples are given to demonstrate how the model is used to design a stepper motor and a robotic rotary joint.

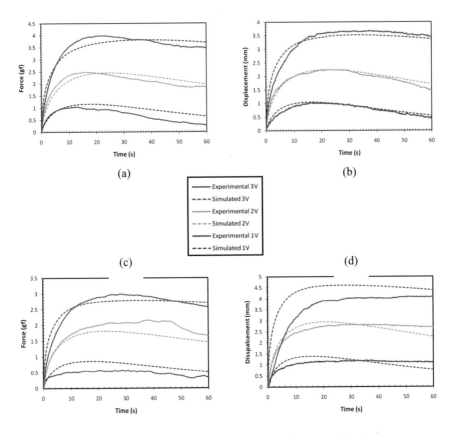

Figure 2.12. *Experimental and simulated (a) blocked force and (b) displacement of a 30 mm long IPMC, and (c) blocked force and (d) displacement of a 35 mm long IPMC [MCD 10b]*

2.3. IPMC stepper motor

A stepper motor has been developed which can operate in air and works by converting the bending actuation of IPMCs into a rotational motion of the motor. The motor is designed with a view to miniaturization and use in micro-robotics where the force output requirements are low, but other advantages of IPMCs are important, such as weight and power availability for remote and embedded systems.

The development of this novel stepper motor demonstrates an innovative mechatronics design process for a complete system with integrated IPMC actuators. The motor has been developed by utilizing the novel model for IPMC actuators incorporated with a complete mechanical model of the motor. The entire system is simulated, and an appropriate size IPMC strip chosen to achieve the required motor specifications and its performance verified. The system has been built and the experimental results are validated to show that the motor works as simulated and can indeed achieve continuous 360° rotation, similar to conventional motors.

2.3.1. *Mechanical design*

The conventional stepper motor is a brushless, synchronous electric motor which divides a full rotation of the motor into a number of "steps". A stepper motor configuration shown in Figure 2.13 was chosen for achieving rotary motion using IPMCs for a number of reasons, including simple design and working mechanism to convert IPMC bending to rotary motion, very low contact area between IPMC and device resulting in low friction, and also the ability to use open loop control architecture. Advantages of the IPMC stepper motor in comparison to a traditional stepper motor include low cost, lightweight design and low power consumption.

The IPMC stepper motor works by sending a voltage sequence to the IPMC actuators which will cause them to bend into contact with the pins attached to the motor shaft; this applies a force to the motor which will then result in controlled rotary "stepping" motion. The pins are placed on the top and bottom layer, each corresponding to one IPMC. The pins on each layer have 90° separation from each other and the top pins are 45° out of phase from the bottom pins. The stand is adjustable for accommodating different size IPMC strips as required. A pulsed input voltage of ±3 V is used to actuate the motor, with each pulse or step corresponding to a 45° rotation.

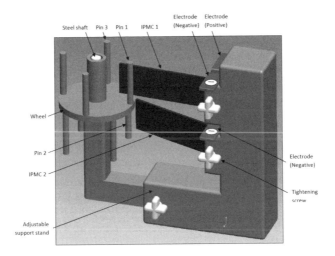

Figure 2.13. *CAD model of the proposed stepper motor [MCD 10a]*

2.3.2. *Model integration and simulation*

The stepper motor has been designed using CAD tools and mathematical analysis and is integrated with the developed IPMC actuator model. The motor friction is added to the system model, using a standard Coulomb and viscous friction model, to make a realistic simulation and see if the IPMC can actually move the motor shaft. The motor friction force was found experimentally to be 0.27 gf for static and 0.21 gf for dynamic motion. Using these forces the friction coefficients were calculated and inputs obtained to create an entire system model which can accurately represent the real life situation.

Different lengths of IPMC were simulated and the length of 35 mm long (clamped length 5 mm) and 10 mm wide was found to give the desired performance, both force and deflection, that was required to actuate the motor. Designing the system in simulation has first allowed the system to be extensively tested and the performance verified before the prototype was built. This demonstrates the usefulness of the developed model for designing IPMC-actuated mechanisms. The simulated performance of the system is shown in Figure 2.14. It can be seen that there is a pause in the operation of the motor between steps. This is necessary in the design with two IPMCs to avoid the motor pins clashing with the IPMC that is returning to its home position.

2.3.3. *Experimental validation*

The actual stepper motor is rapid prototyped, shown in Figure 2.15, and experiments are undertaken to test the actual performance and verify all the simulations. Figure 2.16 shows the actual measured tip displacements of the IPMCs when they are being actuated in order to move the motor. They are measured using 2 Banner LG10A65PU laser sensors with a 3 μm resolution.

There is a reasonable correspondence between the simulated motor and the actual experimental results. It can be seen that at the beginning the IPMC has a larger displacement and it starts to degrade in performance. This is mainly due to the fact that the motor is operating in air for a period of time and the IPMCs exhibit highly time-varying behavior in this type of environment. Despite this the motor does act as the simulation predicts and rotary motion is achieved. The camera shots in Figure 2.17 show the motor in operation, and again it can be seen that the system does operate as predicted.

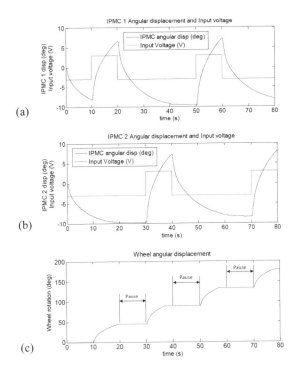

Figure 2.14. *Simulated input voltage and displacement for (a) IPMC 1 and (b) IPMC 2 and (c) resulting motor shaft displacement [MCD 10a]*

Figure 2.15. *Rapid prototyped stepper motor*

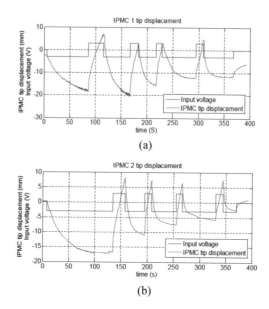

Figure 2.16. *Laser displacement of the IPMCs driving the stepper motor [MCD 10a]*

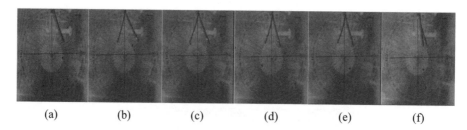

| (a) | (b) | (c) | (d) | (e) | (f) |

Figure 2.17. *Camera shots of motor in operation (a) at initial position, (b) after step 1,
(c) after step 3, (d) after step 5, (e) after step 7, (f) 315° rotation completed and
IPMCs back to home position (Note: the black dot) [MCD 10a]*

2.3.4. *Extension to four IPMCs*

It has been demonstrated that the stepper motor works as simulated and therefore
it is valid to believe that this model and simulation technique can be extended to
other devices. The next step to improving the performance of the stepper motor is to
remove the pause in the operation to achieve continuous rotation as with traditional
rotary motors.

A new design incorporating four IPMCs and using the same principle as the previous design has been proposed. The IPMCs work in two pairs, with each pair acting similarly to the two IPMCs in the previous design. In simulation it is shown that the two pairs must be 90° or more out of phase from each other to avoid the IPMC clashing when returning to the home position. The phase also has to be a multiple of 45° to remove the pause and achieve constant motion. It has therefore been designed that the IPMC pairs are 135° out of phase. The simulation results are shown in Figure 2.18 and prove that this will remove the pause in the system and that the new design can indeed achieve continuous motion.

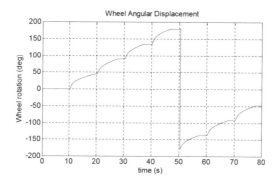

Figure 2.18. *Motor shaft angular displacement for a four IPMC stepper motor [MCD 10a]*

The time step can be altered to speed up the motor, but there is a limit to the step time as the IPMCs still need to have enough time to reach their home positions. A smaller voltage could also be applied, but this would decrease the available torque output of the motor and may decrease the speed of the motor as well. Variations on the design can easily be simulated using the model to verify their design before going on to build the system.

2.4. Robotic rotary joint

A variety of literature has been produced on free-bending of IPMC actuators, but little research has been carried out on using them to drive mechanisms. This application explores the potential for applying IPMC to driving miniature rotary mechanisms, for small-force robotic manipulators, or positioning systems. Rotary joints are commonly found in robotics, industrial applications, biology, and many other areas.

The motivation is to develop a system that can be used in a much wider range of applications that would benefit from lightweight, flexible actuators driving mechanisms such as linkages and rotary joints, which account for a sizeable portion of modern technology. Furthermore, real life issues such as mechanism dynamics, friction, and weight need to be tackled so these actuators could potentially be used as replacements for existing, more bulky devices.

2.4.1. *Mechanical design*

A simple lightweight, rigid, single degree-of-freedom rotary joint has been designed which incorporates an IPMC in the commonly used cantilevered configuration, as shown in Figure 2.19. The rotary linkage has a length of 30.1 mm, a width of 20.0 mm wide, and total weight of 1.1 g. At the end of the rotary linkage is a slot compartment. The IPMC pushes against the side walls of the slot when actuated to move the rotary linkage. When assembled, the rotary linkage can be driven up to 40° in each direction which exceeds the capabilities of the IPMC actuators based on the results of the open loop deflection experiments. Results from 20 consecutive friction tests revealed that the average blocking force of the rotary mechanism is 0.084 gf with a standard deviation of 9.725×10^{-5} gf.

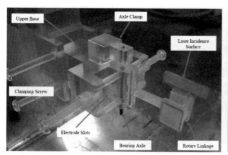 

Figure 2.19. *Photo of the rotary mechanism (a) without IPMC and (b) actuated by an IPMC with –4 V step voltage*

2.4.2. *Control system*

In order to harness the wide-ranging advantages of IPMC actuators and to aid their successful implementation into real systems, the actuation response of an IPMC must be effectively controlled. However, this is not a trivial task due to their complex behavior. Open loop control has been successfully implemented for the stepper motor but more advanced closed loop control will be needed for the rotary joint.

In order to effectively control the system displacement a one degree-of-freedom proportional integral derivative (PID) control architecture is implemented, as shown in Figure 2.20. It has been demonstrated in [RIC 03, YUN 06b, FAN 07, MCD 10c] that simple linear PID control can in fact accurately control the IPMC system over a certain operating range.

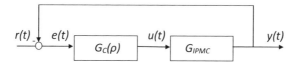

Figure 2.20. *One degree-of-freedom PID closed loop control system*

2.4.3. *System parameter tuning*

Most IPMC controllers that have been realized are tuned in simulation using an approximate plant model. The performance is then assessed, also in simulation, before implementing the controller on the real system. One major issue with this method is the development of a suitably accurate IPMC model which is complex and time consuming as the IPMC is extremely nonlinear, time-variant, and environmentally sensitive. In addition the controller must be further fine-tuned on the real system to account for variability between the model and the real plant. The controller is also sample specific so it cannot be transferred to a different IPMC actuator. Consequently it is highly desirable to develop an automatic tuning method.

Iterative feedback tuning (IFT) is an automatic tuning method which seeks to optimize the controller parameters in the system. The IFT algorithm applied to this application enables the IPMC to be adaptively tuned without the need of any model or knowledge of the system. We simply set the IPMC and then run the tuning algorithm to obtain an optimally tuned system. This controller tuning method eliminates the need for an accurate model in order to accurately control the IPMC actuator. This is a completely new way of thinking for smart materials as up until now research emphasis has been on the modeling of materials in order to be able to control their behavior. This new model-free approach to controller design presents a major step forward for IPMC technology, toward wide acceptance as a viable alternative to traditional actuators.

Even though some model-based control methods have shown reasonable performance they have only been proven to operate well only over a short period of time. The fact is that IPMC dynamics do vary far from their initial state and the performance of a static model-based controller can become unacceptable. If the system dynamics of an IPMC drift far from the developed model then the effort spent on modelling the system becomes redundant. The system dynamics can change so much that they will even shift outside the acceptable range of a robust control design. The IFT algorithm overcomes these issues as it will adaptively tune toward an optimal state whatever the system dynamics change to.

It has been decided that since a model has been previously developed for the design and development of mechanical systems then it will be useful to design a model-based controller for comparison between the performance of the IFT algorithm and a model-based design, demonstrating IFT's adaptive ability in comparison with a non-adaptive model-based control system. The model-based controller can also be used as a starting point for the IFT, in this way the IFT algorithm will take over the model-based approach and automatically tune the system toward an optimal state as the system dynamics change.

2.4.3.1. *IFT overview*

IFT is an automatic tuning method that iteratively optimizes the controller parameters, which are used to regulate the performance of an unknown plant [HJA 94]. This tuning method uses the response of the actual system to determine new updated and improved control parameters in order to minimize some cost function of the system, in this case a least squares fit of the tracking error. As the controller parameter updates are based on only experimental data from the actual system, there is no need for any knowledge or model of the system. The implementation of IFT has shown good results in both laboratory and industrial applications such as control of profile cutting machines [GRA 07], speed and position control of servo drive [KIS 09], temperature regulation in a distillation column [HJA 98], and control of photo resistant film thickness [TAY 06]. The method has traditionally been used to tune systems off-line, or before they commence standard operation due to the fact that a "special" gradient experiment, whose trajectory may deviate far from the trajectory of the normal experiment, is needed to calculate the updated parameters. A full description of the IFT algorithm can be found in [HJA 98] and a concise explanation of the key details for implementation is presented in the following section.

2.4.3.2. *Algorithm for IFT implementation*

The control system presented in Figure 2.20 will be tuned using the IFT algorithm to successfully control the IPMC system. IFT is a time domain approach whose objective is to minimize a cost function or design criterion based on the controller performance in order to obtain an optimally tuned system. There are a number of different design criteria which have been proposed in the literature based on tracking error and control effort. The design criteria that will be used for the rotary joint controller is a quadratic function based on a least squares fit of the tracking error, $\tilde{y}_t$, as shown below in equation [2.15].

$$j(\rho) = \frac{1}{2N} \sum_{t=1}^{N} \left[\tilde{y}_t (\rho)^2 \right]$$

[2.15]

where ρ is a vector of the controller parameters to be tuned, N is the total number of time steps for a given experiment and $\tilde{y}_t(\rho) = y_t(\rho) - r\mathbb{F}$ is the system error at discrete time step t. The premise of the IFT algorithm is to find a minimum of the design criteria, J, in this case finding the minimum tracking error over the entire experiment. In order to locate the minimum of the criteria the gradient is found by differentiating equation [2.15] and finding the solution to make this equal to 0, equation [2.16].

$$\frac{\partial J(\rho)}{\partial \rho} = \frac{1}{N} \sum_{t=1}^{N} \left[\tilde{y}_t (\rho) \frac{\partial y_t (\rho)}{\partial \rho} \right] = 0$$

[2.16]

Applying the iterative algorithm in equation [2.17], the solution for ρ can be found to obtain the minimum error for the system. This is a gradient search algorithm:

$$\rho_{i+1} = \rho_i - \gamma R_i^{-1} \frac{\partial J(\rho_i)}{\partial \rho}$$

[2.17]

where R_i is an appropriate positive definite matrix which determines the search direction for the optimization, i is the iteration number and γ is a positive real scalar which controls the step size. Using the identity matrix for R_i gives a negative gradient direction. It is commonly accepted in the literature [HJA 98, HJA 02, GRA 07] that using the Gauss–Newton approximation of the Hessian of $J(\rho)$ for R_i gives improved results. This becomes more important when the sample size is small. The Hessian is given below in equation [2.18].

$$R_i = \frac{1}{N} \sum_{t=1}^{N} \left(\frac{\partial y_t (\rho_i)}{\partial \rho} \left[\frac{\partial y_t (\rho_i)}{\partial \rho} \right]^T \right)$$

[2.18]

In order to solve for the updated controller, two signals are needed, $\tilde{y}_t(\rho)$ and $\partial y_t(\rho)/\partial \rho$, for equation [2.16]. These must be found independently such that they are unbiased by each other [HJA 02]. In the standard IFT algorithm these signals are found over two independent experiments as follows:

1. A first experiment is conducted under normal operating conditions with an external deterministic reference signal, r, applied at the input, and the output y is recorded. $\tilde{y}_t(\rho)$ can then be found from $\tilde{y}_t(\rho) = y_t(\rho) - r$.

2. A second, "special" experiment is then conducted in order to calculate the gradient, $\partial y_t(\rho)/\partial \rho$. This is the same as the first experiment, except the input, r, is the error $\tilde{y}_t(\rho)$ from the first experiment.

With these two batches of data $\tilde{y}_t(\rho)$ is the error from the first experiment and $\partial y_t(\rho)/\partial \rho$ is calculated by the following.

For the given control system used for tuning the IPMC, Figure 2.20, the closed loop output is defined as:

$$y(\rho_i) = \frac{G_C(\rho_i) G_{IPMC}}{1 + G_C(\rho_i) G_{IPMC}} r \qquad [2.19]$$

Then by differentiating the output, $\partial y_t(\rho)/\partial \rho$ can be found as:

$$\frac{\partial y(\rho_i)}{\partial \rho} = \frac{1}{G_C(\rho_i)} \frac{\partial G_C(\rho_i)}{\partial \rho} \left[\frac{G_C(\rho_i) G_{IPMC}}{1 + G_C(\rho_i) G_{IPMC}} (r - y(\rho_i)) \right] \qquad [2.20]$$

By comparing the term in the square brackets in equation [2.20] and equation [2.19] it can be seen that the term in the square brackets is the result of injecting the error from the first experiment through the closed loop system. The output from the plant for this second experiment gives the term in the square brackets in equation [2.20]. The two terms $\dfrac{1}{G_C(\rho_i)} \dfrac{\partial G_C(\rho_i)}{\partial \rho}$ can be found by differentiating the controller itself and hence $\dfrac{\partial y(\rho_i)}{\partial \rho}$ can be established. Using this result, the Hessian can be calculated from equation [2.18] and also $\dfrac{\partial J(\rho)}{\partial \rho}$ can be found leading to the new updated controller parameters, ρ_{i+1}, which will give an improved controller for the system. This procedure is then repeated for the desired number of iterations, or until the desired system performance is achieved.

2.4.4. *Experimental tuning results*

Experiments were undertaken using a custom test rig which supports 2 copper clamps which act as electrodes to pass the voltage to the IPMC. The IPMC and clamps are placed in a container of de-ionized water in order to avoid rapid dehydration and potential damage to the IPMC. This will also slow the time-varying behavior so the performance of the tuning algorithm can be more objectively assessed. Testing was undertaken based on the free deflection of the IPMC to prove the control system required for the robotic joint.

A Nafion®-based IPMC was used, with Pt electrodes. The IPMC was 35 mm long, 10 mm wide with a thickness of 200 μm. The clamped length was 5 mm. This relatively long length of IPMC was chosen because most research has been carried out with shorter lengths of IPMC as actuation response is more linear with shorter IPMCs [ANT 08, HUN 08]. This research is attempting to tackle the nonlinearity so therefore a long length was used. Also shorter IPMCs cannot achieve large displacements so a long length will be needed to ensure that both micro (<1 mm) and macro (>1 mm) displacements can be investigated. With the specific IPMC used for this research, up to a 3 mm displacement will be input as the target reference.

Due to the desired applications in robotics and biomimetics, the control system was designed to be accurate for changes in set point, in terms of both the transient and steady-state response. In order to tune for this, the reference trajectory was a stair-step function to the desired set point to be tuned. The experiments will be 60 s long with a reference of: positive target displacement for the first 15 s, then step back to zero displacement until 30 s, then negative target displacement until 45 s, and finally to zero displacement until 60 s. This reference will ensure that the IPMC has been tuned in both directions, as it has been shown that due to imperfect fabrication techniques the IPMC can have different performance in different directions. This reference will give tuning for four transient periods as well as steady-state behavior.

The initial controller parameters, K_p, K_i, and K_d, were chosen by tuning the IPMC through model simulations. This will give the benchmark model-based control system for performance comparison. The derivative gain in a PID controller contributes based on the change in error, and therefore will amplify any high frequency noise that may be present in the laser sensor or control electronics. It was desired to control the IPMC to micron displacements, where the noise starts to become an appreciable part of the feedback signal, so a high K_d value is likely to introduce large high frequency oscillation and possibly make the system unstable. Also it has been shown by Liu in 2010 [LUI 10] that PI controllers can exhibit good response in controlling IPMCs. For these reasons the authors were confident to start with a PI controller by setting the derivative gain to 0 and letting the IFT tuning algorithm decide how much derivative action to include. The chosen initial model-based values were $K_p = 1,000$, $K_i = 500$, and $K_d = 0$.

In order to ensure convergence to the local minimum of the design criteria the data set has been chosen large, 600 samples (60 s at 10 Hz), and the step size for the control parameters must be chosen relatively small. The step size must be small enough to ensure the controller does not "jump too far" and result in an unstable system, but be large enough so that there is a rapid convergence to the minimum design criteria, otherwise too many iterations will be needed, making the algorithm impractical.

The step sizes chosen for the IPMC system were $\gamma K_p = 1$, $\gamma K_i = 1$, and $\gamma K_d = 0.5$. As a rule for the IPMC system the step size was chosen so that control parameters would update by no more than 100% of the previous value. From the experiments undertaken it has been shown that this step size will ensure that the system will remain stable, but also achieve a rapid convergence within five iterations. The value for step size of the derivative gain was chosen as half of that for the proportional and integral gain because for a large increase in derivative term it is possible the system may iterate to an unstable system at low deflections in the presence of large noise input.

With the setup completed, the IFT algorithm was run on the IPMC using the initial controller values and step sizes. The IFT algorithm was run to tune the IPMC controller for displacements in the micro- and macro-range, reference signals ranging from 100 µm to 3 mm. This was done as due to the nonlinear nature and varying dynamics of the IPMC at varying displacements it is expected that the optimal controller parameters would vary considerably depending on the target displacement.

The time response for the 3 mm target displacement is shown in Figure 2.21. First the initial output using the model-based PI controller is shown (see Figure 2.21(a), then the performance of the system with the next five consecutive iterations of the controller parameters, found by the automatic IFT algorithm, are shown (see Figures 2.21 (b)–(f) respectively).

It is clear to see that the initial controller had large oscillation in the first quarter, large overshoot starting at 30 s, and also some overshoot and oscillation when finally returning to zero displacement. Even after just one iteration of the control parameters there was an obvious improvement in the response, with the oscillations in the first 15 s reduced significantly as well as the overshoot at 30 s. After each iteration the controller parameters were updated and it can be seen from the time response that the IPMC output drastically improved. After five iterations the performance of the controller for a 3 mm displacement had significantly improved. The improvement can be quantified as a 56% of the design criteria $J(\rho)$.

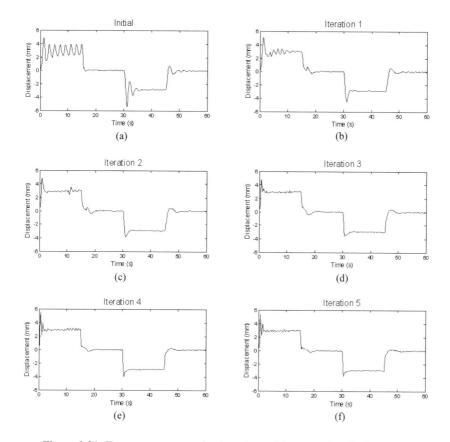

Figure 2.21. *Time response over five iterations of the controller for 3 mm step displacement [MCD 10c]*

This same procedure was undertaken for 100 µm, 200 µm, 300 µm, 500 µm, 1 mm, 1.5 mm, 2 mm, 2.5 mm, and 3 mm displacements. It was found that at 100 µm and 200 µm, despite being able to track accurately to the target displacements, the noise level from the sensor and electronics was too large to accurately tune for these displacements. The tuned parameters were very inconsistent because of the low signal-to-noise ratio (SNR). This was confirmed by the fact that the IFT algorithm kept the proportional gain at these displacements at zero or very low to ensure that noise was not amplified.

The full set of results for the final tuned values for each displacement is given in Table 2.3, along with the percentage improvement of the controller design criterion. The major success of the IFT algorithm in tuning the system from the initial model-based controller is very clearly seen.

Target (mm)	Final K_p	Final K_i	Final K_d	Initial J $(\times 10^{-9})$	Final J $(\times 10^{-9})$	% improvement
0.3	3092.56	2813.93	273.26	3.2411	1.9618	39.47
0.5	2636.35	2791.43	483.04	11.383	3.9523	65.28
1	2019.93	1904.23	436.34	29.289	12.650	56.80
1.5	1480.53	1903.93	476.95	58.403	26.322	54.93
2	1488.18	1585.71	539.94	91.605	52.129	43.09
2.5	1577.95	1128.05	708.01	296.87	113.38	61.81
3	980.449	1040.13	354.75	288.56	127.93	55.66

Table 2.3. *Summary of the results for IFT at different target displacements*

2.4.5. *Gain schedule nonlinear controller*

In order to achieve accurate control over a large range of target displacements, on the micro- and macro-scale, a gain scheduled (GS) nonlinear control architecture is developed to extend the PID control system. This allows the control parameters to adapt depending on the states of the system. This is necessary as it can be seen that the optimal controller gains do vary significantly based on the reference input. This GS controller has the ability to automatically adapt itself in order to tackle the nonlinearities and time-variance to achieve accurate positioning over a large displacement range.

The schedule for the controller gains has been chosen based on a function of the reference trajectory, that is the gains vary depending on the target displacement of the joint. Its architecture is shown in Figure 2.22 and the task now is to find the function $f_{GS}(r)$ for the controller.

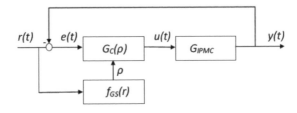

Figure 2.22. *Block diagram of proposed GS controller*

The PID controller has been tuned using IFT for a range of different target values, resulting in a set of tuned linear controllers for different reference trajectories. Now in order to turn the finite linear controllers into a continuous controller, the control parameters must be interpolated in order to schedule the gain continuously over the operating range. The tuned controller gains K_p, K_i, and K_d are plotted in Figure 2.23 as a function of target displacement. There is a clear trend in the relationship with the control parameters and target displacements. After some analysis it was found that a logarithmic fit would give the best correlation between these parameters. These trends for the control parameters are plotted on the graph in Figure 2.23 and the relationships are presented in equations [2.21]–[2.23].

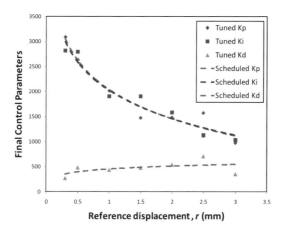

Figure 2.23. *Final controller parameters after IFT at varying displacements [MCD 10c]*

$$K_p = -839.7 \ln(r) - 3757.6 \qquad [2.21]$$

$$K_i = -808.0 \ln(r) - 3560.2 \qquad [2.22]$$

$$K_d = 84.73 \ln(r) + 1038.1 \qquad [2.23]$$

The controller was also tuned for references of 100 μm and 200 μm, but because of the low SNR the tuned parameters were very inconsistent. The tuned values for K_p and K_i at 100 μm and 200 μm were in the region of those obtained for 300 μm and the values for K_d were slightly lower than that for 300 μm at these small targets. For this reason it was decided to restrict the controller parameters at their 300 μm values, that is if the target displacement was less than 300 μm then use the values scheduled for 300 μm. This is also necessary to prevent K_p and K_i from increasing extremely high because their schedule approaches the zero reference asymptotically.

An interesting observation is that if the IPMC was linear, then the controller gains would be constant across all target displacements. This in itself validates that the IFT algorithm has successfully tackled the nonlinear characteristics of the IPMC. It can also be observed that the IFT algorithm realizes that the SNR is low at small displacements and consequently reduces K_P accordingly as not to amplify the noise at these levels.

Now the schedule to change the gains, $f_{GS}(r)$, has been developed, the nonlinear GS controller is complete and ready for implementation.

2.4.6. *Gain schedule vs. PID controller*

The developed nonlinear GS controller was tested and its performance compared to a conventional PID controller with the IFT-tuned parameters for 1.5 mm displacement, as this is in the middle of the range of IPMC operation. The design criteria for IFT, equation [2.15], was used as a quantitative measure of the performance of the controllers and qualitative measures such as overshoot and settling time are also analyzed.

In order to test the performance of the GS controller versus the conventional PID controller for changes in set point, which is what the controller is designed for, a random stair-step sequence of varying amplitude was used as input. Both the micro and macro targets were used. The results of this test are shown in Figure 2.24. By examining the plot it can be seen that the GS controller had a much smaller overshoot at all of the set point changes, except for at micro targets (30 s and 90 s). This is due to the fact that the proportional gain is high when the target displacement is low as seen in Figure 2.23. This suggests that the cut-off region in the schedule, which was set at 300 μm, may be too low. Using a higher cut-off (say 500 μm) will restrict the K_p and K_i values at micro displacements and the GS controller may not over shoot as much.

The settling time after a set point change is better for the GS controller in all cases, even at the micro targets (30 s and 90 s), where there is more overshoot. Comparing the overall error for the controllers using the design criteria the GS controller is 17% better.

$$J(\mathrm{PID}) = 1.19e^{-7}$$

$$J(\mathrm{GS}) = 1.02e^{-7}$$

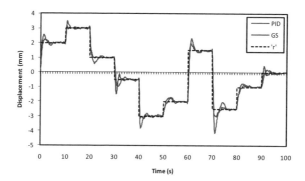

Figure 2.24. *Random stair-step reference input for comparison of control performance [MCD 10c]*.

In order to demonstrate the versatility of the GS controller to other input signals, sinusoid reference trajectories were tested versus the conventional PID controller. This will result in dynamically varying control parameters as the reference signal is continuously changing. A number of experiments were undertaken to assess the performance under different conditions inside the desired operating range of the IPMC.

Figure 2.25(a) shows the 33×10^{-3} Hz signal with a micro amplitude of 500 μm. It can be seen that both controllers follow the reference very well, despite a relatively high level of noise. By inspection, the performance of both controllers are comparable, but using the design criteria it can be seen that the GS controller does perform slightly better, $J(\text{PID}) = 1.32e^{-9}$ and $J(\text{GS}) = 1.15e^{-9}$.

Figure 2.25(b) shows the 33×10^{-3} Hz signal with a large amplitude of 3 mm. Both controllers track the reference extremely well and again by inspection the performance of both controllers are again comparable. Using the design criteria, $J(\text{PID}) = 6.55e^{-9}$ and $J(\text{GS}) = 5.35e^{-9}$, so again the GS controller does perform better.

It has been shown that the designed controllers can accurately track a dynamic reference with a time period of 30 s, so a faster signal of 0.1 Hz was tested. Figure 2.26(a) shows the 0.1 Hz performance at a 500 μm amplitude for the two controllers. It can be seen that the standard PID controller has a considerable level of overshoot and then consequently lags the reference signal. It is clear that the GS controller is performing better and this is confirmed by the design criteria, $J(\text{PID}) = 16.3e^{-9}$ and $J(\text{GS}) = 2.35e^{-9}$.

Figure 2.26(b) shows the performance with a reference amplitude of 3 mm at 0.1 *Hz*. Similar to the 500 μm reference it is clear to see that the standard PID controller exhibits overshoot. This high level of overshoot again results in an output lag. By inspection the GS controller performs a great deal better and this can be confirmed by the design criteria, $J(\text{PID}) = 13.6e^{-8}$ and $J(\text{GS}) = 6.96e^{-8}$.

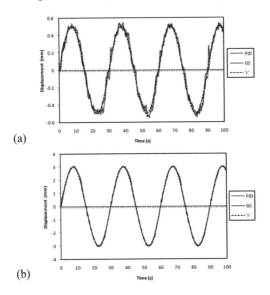

(a)

(b)

Figure 2.25. 33×10^{-3} Hz sinusoid inputs for (a) 500 µm and (b) 3 mm amplitude [MCD 10c]

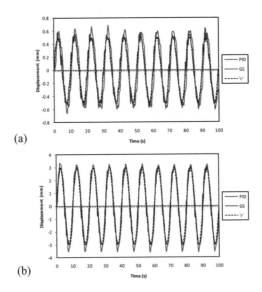

(a)

(b)

Figure 2.26. 0.1 Hz sinusoid inputs for (a) 500 µm and (b) 3 mm amplitude [MCD 10c]

2.5. Discussions

Although modeling the IPMC has been a large research focus for 15 years there is still no complete and widely accepted model that has been developed to predict the mechanical output and even the underlying mechanisms for actuation are not fully understood. This is partly due to the very complex nonlinear, time-varying, and environmentally sensitive nature of the composite material [YAG 04, BON 07, BUF 08]. The development of a suitable IPMC model which will accurately represent the real system is extremely complex and time consuming. The model presented acts as a very useful tool for mechanical design and simulation of the mechatronics systems. It however is not accurate enough over time to develop a controller based solely on this or any other model. It has however served as an excellent tool for mechanical design and simulation for the applications presented and has given a good starting point to develop an acceptable stable controller, for the IFT algorithm to tune from.

The IFT algorithm has the major advantage of being model-free which is extremely useful for IPMCs and any other mechatronics systems where a model of the system is not available and developing one maybe too complex. With regard to IPMCs this is a major step forward, as modeling has been a large research focus for 15 years and there is still no widely accepted model for the actuation response. This will have a major impact in this field and will aid in the implementation of IPMCs into real mechatronics systems, which up until now have been restricted due to limitations in modeling and control.

The hysteresis behavior and other nonlinearities are not directly addressed by the IFT controllers, as some model or knowledge of the system would be necessary to account for this [ZHE 08], which is specifically what the IFT algorithm is avoiding. Despite this the controller may well instinctively compensate for some part of the hysteresis as it is automatically tuning online as the system operates in certain modes.

The IFT tuning has a modified experiment every third time period, even in the online case, which can cause some deviation from the desired reference. To minimize the effect of this the tuning experiment may be undertaken say every 10th or even 100th iteration, depending on how quickly the system is varying and how quickly the system needs to converge to an optimal state. For example if a system is operating all day, it may need to be tuned for an hour per day to keep the system operating successfully.

2.6. Concluding remarks

Two mechatronic devices have been developed through a thorough design process. First the IPMC actuator was modeled in order to develop a useful tool for developing and testing the performance of the applications before implementing them in real life. Control systems have been developed and implemented using novel approaches, each tailored to the specific application; i.e. open loop control for the stepper motor, PID, and GS control for the rotary joint mechanism. This has successfully demonstrated the capabilities of IPMCs and the benefits of implementing them as valid alternatives to traditional mechatronic actuators.

2.7. Bibliography

[AHN 10] AHN K.K. *et al.*, "Position control of ionic polymer metal composite actuator using quantitative feedback theory", *Sensors and Actuators A: Physical*, vol. 159, 2010, pp. 204–212.

[ANT 08] ANTON M. *et al.*, "A mechanical model of a non-uniform ionomeric polymer metal composite actuator", *Smart Materials and Structures*, vol. 17, no. 2, 2008, p. 025004.

[BAR 00a] BAR-COHEN Y. *et al.*, "Challenges to the application of IPMC as actuators of planetary mechanisms", *Proceedings of SPIE – The International Society for Optical Engineering*, vol. 3987, 2000, pp. 140–146.

[BAR 00b] BAR-COHEN Y. *et al.*, "Challenges to the transition to the practical application of IPMC as artificial-muscle actuators", *Materials Research Society Symposium – Proceedings*, 2000.

[BON 07] BONOMO C. *et al.*, "A nonlinear model for ionic polymer metal composites as actuators", *Smart Materials and Structures*, vol. 16, no. 1, 2007, pp. 1–12.

[BUF 08] BUFALO G.D. *et al.*, "A mixture theory framework for modeling the mechanical actuation of ionic polymer metal composites", *Smart Materials and Structures*, vol. 17, no. 4, 2008.

[CHE 09] CHEN Z. *et al.*, "A nonlinear, control-oriented model for ionic polymer-metal composite actuators", *Smart Materials and Structures*, vol. 18, no. 5, 2009.

[CHE 08] CHEN Z., TAN X., "A control-oriented and physics-based model for ionic polymer-metal composite actuators", *IEEE/ASME Transactions on Mechatronics*, vol. 13, no. 5, 2008, pp. 519–529.

[CHE 09] CHEW X.J. *et al.*, "Characterisation of ionic polymer metallic composites as sensors in robotic finger joints", *International Journal of Biomechatronics and Biomedical Robotics*, vol. 2, no. 1, 2009, pp. 37–43.

[FAN 07] FANG B.K. *et al.*, "A new approach to develop ionic polymer-metal composites (IPMC) actuator: Fabrication and control for active catheter systems", *Sensors and Actuators A: Physical*, vol. 137, no. 2, 2007, pp. 321–329.

[GRA 07] GRAHAM A.E. *et al.*, "Rapid tuning of controllers by IFT for profile cutting machines", *Mechatronics*, vol. 17, 2007, pp. 121–128.

[HJA 02] HJALMARSSON H., "Iterative feedback tuning – an overview", *International Journal of Adaptive Control and Signal Processing*, vol. 16, no. 5, 2002, pp. 373–395.

[HJA 94] HJALMARSSON H. *et al.*, "A convergent iterative restricted complexity control design scheme", *Proceedings of the 33rd IEEE Conference on Decision and Control*, 14–16 December 1994.

[HJA 98] HJALMARSSON H. *et al.*, "Iterative feedback tuning-theory and applications", *IEEE Control Systems*, vol. 26, 1998, p. 41.

[HUN 08] HUNT A. *et al.*, "A multilink manipulator with IPMC joints", *Proceedings of SPIE – The International Society for Optical Engineering*, 2008.

[KAN 96] KANNO R. *et al.*, "Linear approximate dynamic model of ICPF (ionic conducting polymer gel film) actuator", *Proceedings of Robotics and Automation, IEEE International Conference*, 22–28 April 1996.

[KIS 09] KISSLING S. *et al.*, "Application of iterative feedback tuning (IFT) to speed and position control of a servo drive", *Control Engineering Practice,* vol. 17, no. 7, 2009, pp. 834–840.

[KOT 08] KOTHERA C.S. *et al.*, "Characterization and modeling of the nonlinear response of ionic polymer actuators", *Journal of Vibration and Control*, vol. 14, no. 8, 2008, pp. 1151–1173.

[LAV 05] LAVU B.C. *et al.*, "Adaptive intelligent control of ionic polymer-metal composites", *Smart Materials and Structures*, vol. 14, no. 4, 2005, pp. 466–474.

[LUI 10] LIU D., "Design and control of an IPMC actuated single degree-of-freedom rotary joint, mechatronics engineering", *The University of Auckland*, New Zealand, 2010.

[MAL 01] MALLAVARAPU K. *et al.*, "Feedback control of the bending response of ionic polymer–metal composite actuators", *Smart Structures and Materials 2001: Electroactive Polymer Actuators and Devices*, Newport Beach, CA, USA, SPIE, 2001.

[MCD 09] MCDAID A.J. *et al.*, "A nonlinear scalable model for designing ionic polymer–metal composite actuator systems", *2nd International Conference on Smart Materials and Nanotechnology in Engineering*, WeiHai, China, 2009.

[MCD 10a] MCDAID A.J. *et al.*, "Development of an ionic polymer–metal composite stepper motor using a novel actuator model," *International Journal of Smart and Nano Materials,* vol. 1, 2010, pp. 261–277.

[MCD 10b] MCDAID A.J. *et al.*, "A conclusive scalable model for the complete actuation response for IPMC transducers", *Smart Materials and Structures*, vol. 19, no. 7, 2010, p. 075011.

[MCD 10c] MCDAID A.J. *et al.*, "Gain scheduled control of IPMC actuators with 'model-free' iterative feedback tuning", *Sensors and Actuators A: Physical*, vol. 164, no. 1–2, 2010, pp. 137–147.

[NEM 00] NEMAT-NASSER S., LI J.Y., "Electromechanical response of ionic polymer–metal composites", *Journal of Applied Physics*, vol. 87, no. 7, 2000, pp. 3321–3331.

[NEW 02] NEWBURY K.M., Characterization, modeling, and control of ionic polymer transducers, PhD, Mechanical Engineering, Virginia Polytechnic Institute and State University, 2002.

[POR 08] PORFIRI M., "Charge dynamics in ionic polymer metal composites", *Applied Physics*, vol. 104, no. 10, 2008.

[PUN 07] PUNNING A. *et al.*, "Surface resistance experiments with IPMC sensors and actuators", *Sensors and Actuators A: Physical*, vol. 133, no. 1, 2007, pp. 200–209.

[RIC 03] RICHARDSON R.C. *et al.*, "Control of ionic polymer metal composites", *IEEE/ASME Transactions on Mechatronics*, vol. 8, no. 2, 2003, pp. 245–253.

[SAG 92] SADEGHIPOUR K. *et al.* "Development of a novel electrochemically active membrane and 'smart' material based vibration sensor/damper", *Smart Materials and Structures*, vol. 1, 1992, pp. 172–179.

[SAN 10] SANTOS J. *et al.*, "Ionic polymer–metal composite material as a diaphragm for micropump devices", *Sensors and Actuators A: Physical*, vol. 161, no. 1–2, 2010, pp. 225–233.

[SHA 99] SHAHINPOOR M., "Electromechanics of ionoelastic beams as electrically controllable artificial muscles. Smart Structures and Materials", *Electroactive Polymer Actuators and Devices*, Newport Beach, CA, USA, SPIE, 1999.

[SHA 01a] SHAHINPOOR M., KIM K.J., "Design, development and testing of a multi-fingered heart compression/assist device equipped with IPMC artificial muscles", *Proceedings of SPIE - The International Society for Optical Engineering*, 2001.

[SHA 01b] SHAHINPOOR M., KIM K.J., "Ionic polymer-metal composites: I. Fundamentals", *Smart Materials and Structures*, vol. 10, no. 4, 2001, pp. 819–833.

[SHA 07] SHAHINPOOR M. *et al.*, *Artificial Muscles: Applications of Advanced Polymeric Nanocomposites*, New York, Taylor & Francis, 2007, pp. 119–220.

[TAD 00] TADOKORO S. *et al.*, "Modeling of Nafion-Pt composite actuators (ICPF) by ionic motion", *Smart Structures and Materials 2000: Electroactive Polymer Actuators and Devices (EAPAD)*, Newport Beach, CA, USA, SPIE, 2000.

[TAK 06] TAKAGI K. *et al.*, "Development of a Rajiform Swimming Robot using ionic polymer artificial muscles", *IEEE/RSJ International Conference on Intelligent Robots and Systems*, 2006.

[TAY 06] TAY A. *et al.*, "Control of photoresist film thickness: Iterative feedback tuning approach", *Computers & Chemical Engineering*, vol. 30, no. 3, 2006, pp. 572–579.

[YAG 04] YAGASAKI K., TAMAGAWA H., "Experimental estimate of viscoelastic properties for ionic polymer-metal composites", *Physical Review E*, vol. 70, no. 5, 2004, p. 052801.

[YE 08] YE X. *et al.*, "Design and realization of a remote control centimeter–scale robotic fish", *IEEE/ASME International Conference on Advanced Intelligent Mechatronics*, AIM, 2008.

[YUN 06a] YUN K., A novel three–finger IPMC gripper for microscale applications, PhD, Mechanical Engineering, Texas A&M University, 2006.

[YUN 06b] YUN K., KIM W.J., "Microscale position control of an electroactive polymer using an anti-windup scheme", *Smart Materials and Structures*, vol. 15, no. 4, 2006, pp. 924–930.

[ZHE 05] ZHENG C. *et al.*, "Quasi-static positioning of ionic polymer-metal composite (IPMC) actuators", *Proceedings, 2005 IEEE/ASME International Conference on Advanced Intelligent Mechatronics*, 24–28 July 2005.

[ZHE 08] ZHEN C. *et al.*, "Modeling and control with hysteresis and creep of ionic polymer-metal composite (IPMC) actuators", *Control and Decision Conference, CCDC 2008*, China, 2008.

Chapter 3

Modeling and Simulation of Analog Angular Sensors for Manufacturing Purposes

This chapter develops a new mathematical model, for pancake resolvers, which depends on a set of variables controlled by the sensor manufacturer – the winding parameters. This model allows a resolver manufacturer to manipulate in-process controllable variables in order to readjust already assembled resolvers that without any action would be scrap for the production line. The developed model follows a two-step strategy where on a first step a traditional transformer's model computes the resolver nominal parameters and on a second step a linear model computes the corrections on the controllable variables, in order to compensate for small deviations in design assumptions, caused by the variability of the manufacturing process. At the end of the chapter an experimental methodology for parameter identification and several tests for model validation are presented.

3.1. Introduction

Resolvers are nowadays widely used in industrial applications. These electromagnetic devices which were, in the past, largely used in military applications, namely to control the position stability of heavy guns, are presently very common in industrial areas as a servomotor component. Servomotors are today widely used in robotics, rotary machinery, radars, aeronautics, etc.

The main factors that promote the widespread use of synchros and resolvers as angular sensors are their robustness and stability in non-friendly environments such

Chapter written by João FIGUEIREDO.

as mechanical vibrations, shocks, environments with dust, oil, and radiation. Besides, these electromagnetic devices have very stable properties when subjected to extreme temperature variations (–50° C to +150° C) and high rotational velocities (1,000–10,000 rpm).

The wide application fields of synchros and resolvers can be grouped into three main areas [GOL 81]:

– distant transmission of absolute angles – Figure 3.1;

– analog computation of the difference between two angles (reference and actual values) – Figure 3.2;

– servo-systems (where the information signals are apart from the energy signals) – Figure 3.3.

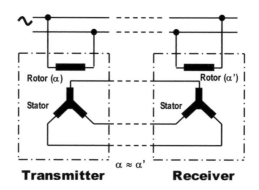

Figure 3.1. *Distant transmission of absolute angles*

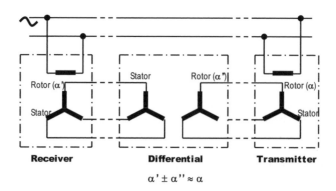

Figure 3.2. *Analog computation of the difference between two angles*

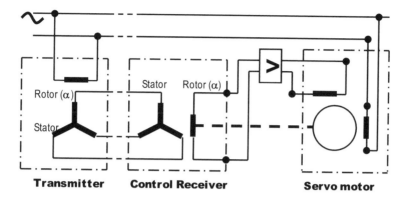

Figure 3.3. *Servo-systems*

The main disadvantages of synchros and resolvers in relation to optoelectronic encoders are both the necessity of an AC-power source and the delivery of an analog output signal. However this latter disadvantage is nowadays vanishing due to the advances in signal processing technology that is constantly delivering more speedy and cost-efficient solutions to convert analog to digital signals. The market's continuous search for more accurate devices and the wide availability of digital controllers coupled to servomotors is increasing today's demand for resolvers as function systems, able to be connected with analog/digital converters as is shown in Figure 3.4.

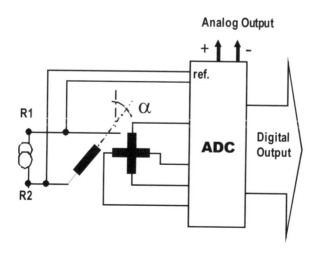

Figure 3.4. *Resolver as a function system coupled to ADC device*

The main research areas focusing resolver development can be grouped into three fields:

– Model development [SUN 08], [BRA 09], [BEN 07], [KIM 09], [HOS 07], [FIG 04], [ALH 04];

– Analog to digital converters [ATT 07], [HAN 90], [VAZ 10], [BRA 08], [SAR 08], [BEN 04], [BEN 05];

– Applications [BUR 08], [BUN 04], [MUR 02], [MAS 00].

This chapter concerns the research field of model development. In this field some authors focus the efforts on improving the model accuracy [SUN 08], [BRA 09], [BEN 07], others concentrate on developing new procedures for modeling and calibration [KIM 09], [HOS 07], [FIG 04].

In [FIG 04] an explicit mathematical model for a pancake resolver is presented. This former paper develops an explicit model for pancake resolvers following the traditional approach from transformer models, where the main parameters are the resistances of primary and secondary coils, the magnetic resistances, and the leakage impedances.

The model developed in [FIG 04] has little use for a resolver manufacturer because it does not give any help supplying a specific path where the manufacturer can act to compensate the deviations in assembled products due to the variability of the manufacturing process.

This chapter enlarges the applicability of the study formerly developed in [FIG 04] by proposing a new mathematical model developed for resolver manufacturers, where the model parameters are the production controllable variables.

The main functional characteristics of a resolver are the angular error, the output voltage (transformer ratio, r), the phase shift, and the input current. All these important factors, specified by applications, are strongly influenced by constructive factors such as magnetic properties of stators and rotors, winding geometries, and manufacturing tolerances of mechanical parts.

Facing this situation, it is clear that the availability of a mathematical model, at the resolver manufacturer, allowing the producer to simulate the characteristics of its products, in the presence of high variability of production factors, is a valuable asset. The availability of a simulation model supplying the resolver production parameters, with high accuracy, implies a smaller number of prototypes needed until customer specifications are effectively met. In the end, this efficiency increase corresponds to a large amount of money that is saved yearly, by the resolver manufacturer, as the number of wasted prototypes is reduced.

The present market for analog angular sensors is extremely competitive as new applications are continuously arriving, encouraging a fall of the prices as a way to reach other markets (the automotive market is today still a marginal market for resolvers but the pressure to expand into this market is extremely high) [MUR 02], [BER 03], [MAP 10].

This chapter develops a new mathematical model for pancake resolvers, dependent on a set of variables controlled by a resolver manufacturer – the winding parameters. The developed model follows a two-step strategy where in a first step a traditional transformer's model computes the resolver nominal conditions and in a second step a linear model computes the corrections on the controllable variables, in order to compensate for small deviations in design assumptions, caused by the variability of the manufacturing process.

3.2. Pancake resolver model

3.2.1. *Description*

The pancake resolver is the most popular resolver in industrial applications and aeronautics because its design avoids the traditional collector that brings energy to the rotor.

The pancake resolver carries the current into the rotor through a transformer that is located at the stator edge. The advantage of such a design, over the traditional resolver with collector, is the absence of the relative movement between mechanical parts which causes wear, vibrations, and sound. Figure 3.5 presents the two above-mentioned designs.

Figure 3.5. *Traditional resolver (left) and pancake resolver (right)*

Independently of how the energy is brought into the resolver rotor, the function of a resolver, as an angular sensor, can be briefly illustrated in Figures 3.6 and 3.7.

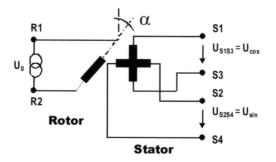

Figure 3.6. *Resolver function schematics*

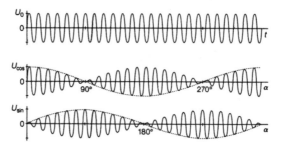

Figure 3.7. *Resolver input and output voltages*

As can be seen, the secondary coils, which are geometrically shifted by 90°, deliver two output voltages that are modulated in magnitude according to the cos and sin functions of the relative angular position of rotor to stator.

This physical effect was first developed by Werner Siemens in 1896 [SIE 96]. Today, the currently available resolvers have an average accuracy of ±0.2°. The angular position of the rotor referred to the stator can be obtained as in [3.1]:

$$\alpha = \tan^{-1}\left[\frac{U_{S2S4}}{U_{S1S3}}\right] = \tan^{-1}\left[\frac{rU_0\sin\alpha}{rU_0\cos\alpha}\right]$$ [3.1]

r = transformer ratio;
α = relative angle rotor to stator;
U_0 = input voltage;
U_{S1S3}, U_{S2S4} = output voltage from each stator winding.

3.2.2. *Mathematical model*

The commonly used mathematical models for resolvers are the typical transformer models that are shown in Figures 3.8, 3.9, and 3.10. These models are very suitable to supply the usual customer electrical characteristics for resolvers, namely the rotor and stator impedances (open and short circuited).

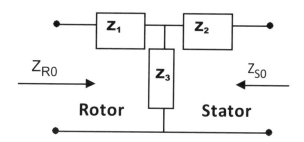

Figure 3.8. *Rotor impedance with stator open (Z_{ro}) and rotor open (Z_{so})*

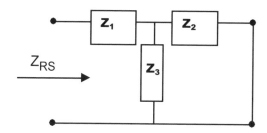

Figure 3.9. *Rotor impedance with stator shorted (Z_{rs})*

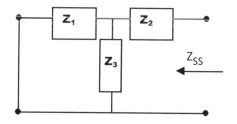

Figure 3.10. *Stator impedance with rotor shorted (Z_{ss})*

According to the below schematics (Figures 3.8–3.10) using the traditional circuit analysis methods, the following equations are derived:

$$Z_{ro} = Z_1 + Z_3 \qquad\qquad [3.2]$$

$$Z_{so} = Z_2 + Z_3 \qquad\qquad [3.3]$$

$$Z_{rs} = Z_1 + \frac{Z_2 Z_3}{Z_2 + Z_3} \qquad\qquad [3.4]$$

$$Z_{ss} = Z_2 + \frac{Z_1 Z_3}{Z_1 + Z_3} \qquad\qquad [3.5]$$

where,

Z_{ro} = rotor impedance with stator open;

Z_{so} = stator impedance with rotor open;

Z_{rs} = rotor impedance with stator shorted;

Z_{ss} = stator impedance with rotor shorted.

This traditional, well-known model, although very useful for computing the main electrical characteristics for customers, is of little use to the manufacturers as it cannot deliver data for production purposes. Actually this model does not supply any controllable parameter for the resolver manufacturer (number of windings and wire diameters).

The new model proposed in this chapter is appropriate for resolver manufacturers because it deals explicitly with the actually controllable variables in a resolver production plant – the winding parameters.

The main resolver variables that directly influence the customer-specific electrical characteristics can be grouped into three areas:

Group 1: *material-related variables* – magnetic permeability of materials (rotor, stator, rotor–transformer, stator–transformer).

Group 2: *geometry-related variables* – dimensional tolerances of mechanical parts (rotor, stator, rotor/stator air-gap, transformer air-gap).

Group 3: *winding-related variables* – spatial distribution of windings, number of windings, and winding wire diameters (in all four components: rotor, stator, rotor–transformer, and stator–transformer).

From these three groups of variables, the resolver manufacturer can directly influence only the third group (winding-related variables). In fact, the other two variable groups are usually fixed for the assembly line as the resolver manufacturer usually buys the materials and parts from external suppliers.

In such a scenario a useful mathematical model for a resolver manufacturer must deal explicitly with the controllable variables at the assembly line – winding-related variables (Group 3).

In Figure 3.11 the geometry and function schematics of a pancake resolver are shown. This product is mainly composed by four main coils, identified as A, B, C, and D (A = stator transformer; B = rotor transformer; C = rotor sensor; D = stator sensor).

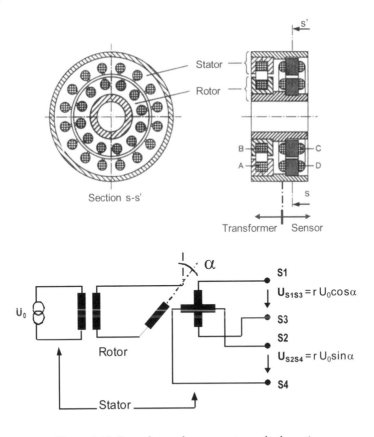

Figure 3.11. *Pancake resolver geometry and schematics*

The study developed in this chapter departs from a traditional transformer model [FIG 04] with design variables (ohmic resistors from primary and secondary coils, magnetic resistances and leakage impedances) and updates this model with a sensitivity model that correlates the main electric specifications of a Resolver with the production factors that are controlled by the manufacturer (mainly winding parameters).

The approach developed in this chapter is inspired by the mathematical methodology of a function expansion according to the Taylor series. The strategy adopted here considers the nominal model developed in [FIG 04] to compute the system nominal values – $f(x_0)$ – and additionally a linear model, dependent on production controllable parameters, which computes the variable increments. These increments will cancel the deviations on the functional characteristics of the resolver, due to the variability of the production processes in the assembly line. The incremental model that is developed in this chapter is an innovative approach based on experimental parameter identification.

3.2.2.1. Model for nominal conditions – [f(x₀)]

Observing the pancake resolver functionality we can consider this device as two standard transformers connected in cascade. The first one supplies the energy into the rotor, with an output voltage that is independent from the relative position of rotor to stator, and the second one, that models the resolver function itself, which can be considered as a rotational transformer with an output voltage that is dependent on the rotor/stator relative angular position.

The model adopted for each one of the above referred transformers is a complete model for a mono-phase transformer, considering magnetic losses in metal and windings. The respective block diagram is presented in Figure 3.12.

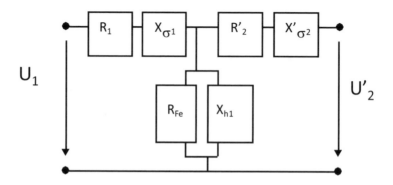

Figure 3.12. *Complete mono-phase transformer model*

The variables represented in Figure 3.12 account for the following effects:

R_1 = primary winding resistance;

R'_2 = affected secondary winding resistance (viewed by the primary winding);

R_{Fe} = magnetic metal resistance;

$X_{\sigma1}$ = primary winding leakage impedance;

$X'_{\sigma2}$ = affected secondary winding leakage impedance (viewed by the primary winding);

X_{h1} = impedance related to common flux.

Connecting two of the above transformer models in cascade we get the complete model for our pancake resolver, which is presented in Figure 3.13. The indexes T and D account respectively for transformer and sensor.

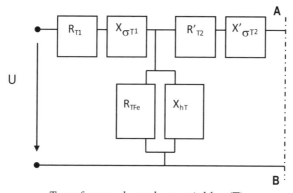

Transformer-dependent variables (T)

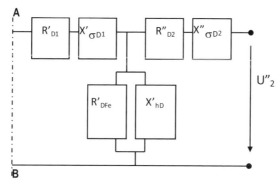

Sensor-dependent variables (D)

Figure 3.13. *Complete pancake resolver model*

Considering the resolver model described in Figure 3.13, we can simplify this model by combining the impedances (serial and parallel), according to the following schematics.

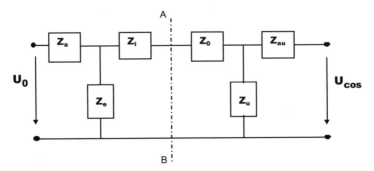

Figure 3.14. *Pancake resolver model – first step simplification*

The simplified impedances are calculated according to the following expressions:

$$Z_a = R_{T1} + X_{\sigma T1} \tag{3.6}$$

$$Z_e = \frac{R_{TFe} \times X_{hT}}{R_{TFe} + X_{hT}} \tag{3.7}$$

$$Z_i = R'_{T2} + X'_{\sigma T2} \tag{3.8}$$

$$Z_0 = R_{D1} + X_{\sigma D1} \tag{3.9}$$

$$Z_u = \frac{R_{DFe} \times X_{hD}}{R_{DFe} + X_{hD}} \tag{3.10}$$

$$Z_{au} = R'_{D2} + X'_{\sigma D2} \tag{3.11}$$

Simplifying the serial association of the Z_i and Z_0 impedances, we obtain the triangular system of impedances illustrated in Figure 3.15.

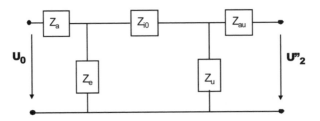

Figure 3.15. *Pancake resolver model – second step simplification*

The simplified impedance Z_{i0} is calculated according to the expression:

$$Z_{i0} = Z_i + Z_0 = R'_{T2} + X'_{\sigma T2} + R'_{D1} + X'_{\sigma D1}$$ [3.12]

The above model can now be reworked in order to present a star topology, as it is illustrated in Figure 3.16.

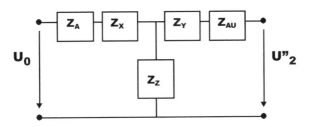

Figure 3.16. *Pancake resolver model – third step simplification*

The simplified impedances (Z_x, Z_y, and Z_z) are calculated according to the following expressions:

$$Z_x = \frac{Z_{i0} \times Z_e}{Z_{i0} + Z_e + Z_u}$$ [3.13]

$$Z_y = \frac{Z_{i0} \times Z_u}{Z_{i0} + Z_e + Z_u}$$ [3.14]

$$Z_z = \frac{Z_e \times Z_u}{Z_{i0} + Z_e + Z_u}$$ [3.15]

After simple manipulations we can reach the pancake resolver simplified model, as it is illustrated in Figure 3.17.

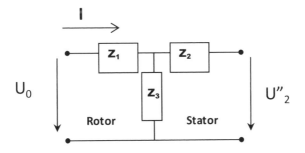

Figure 3.17. *Equivalent-impedance pancake resolver model*

The simplified impedances (Z_1, Z_2, and Z_3) are calculated according to the following expressions:

$$Z_1 = Z_a + Z_x = Z_a + \frac{Z_{i0} \times Z_e}{Z_{i0} + Z_e + Z_u} \qquad [3.16]$$

$$Z_2 = Z_{au} + Z_y = Z_{au} + \frac{Z_{i0} \times Z_u}{Z_{i0} + Z_e + Z_u} \qquad [3.17]$$

$$Z_3 = Z_z = \frac{Z_e \times Z_u}{Z_{i0} + Z_e + Z_u} \qquad [3.18]$$

These impedances can now be written according to the resolver electrical characteristics, using the former equations [3.6] to [3.11] (see Figure 3.13), which gives:

$$Z_1 = R_{T1} + X_{\sigma T1} + \left[\frac{\left(R'_{T2} + X'_{\sigma T2} + R'_{D1} + X'_{\sigma D1} \right)}{Z_{Den}} \right.$$

$$\left. \times \frac{\left(R_{TFe} \times X_{hT} / \left(R_{TFe} + X_{hT} \right) \right)}{Z_{Den}} \right] \qquad [3.19]$$

$$Z_2 = R''_{D2} + X''_{\sigma D2} + \left[\frac{\left(R'_{T2} + X'_{\sigma T2} + R'_{D1} + X'_{\sigma D1} \right)}{Z_{Den}} \right.$$

$$\left. \times \frac{\left(R'_{DFe} \times X'_{hD} / \left(R'_{DFe} + X'_{hD} \right) \right)}{Z_{Den}} \right] \qquad [3.20]$$

$$Z_3 = \frac{\left(\left(R_{TFe} \times X_{hT} \right) / \left(R_{TFe} + X_{hT} \right) \right) \times \left(\left(R'_{DFe} \times X'_{hD} \right) / \left(R'_{DFe} + X'_{hD} \right) \right)}{Z_{Den}} \qquad [3.21]$$

where:

$$Z_{Den} = \left(R'_{T2} + X'_{\sigma T2} + R'_{D1} + X'_{\sigma D1} \right) + \left(R_{TFe} \times X_{hT} / \left(R_{TFe} + X_{hT} \right) \right)$$

$$+ \left(\left(R'_{DFe} \times X'_{hD} \right) / \left(R'_{DFe} + X'_{hD} \right) \right)$$

For industrial applications, beyond the angle measurement accuracy, which was referred to in equation [3.1], the main resolver electrical requirements are:

(i) output voltage from each stator–winding (u_{cos}, u_{sin});

(ii) resolver input current (i).

Considering the representation of dynamic systems according to the transfer function methodology, this chapter computes the resolver mathematical model illustrated in Figure 3.17 and provides the two explicit transfer functions: $U_{cos}(s)/U_0(s)$ and $I(s)/U_0(s)$.

To calculate the transfer functions referred to above we started from the resolver model presented in Figure 3.17. This model is then reworked in order to obtain a frequency model, which is explicitly dependent on the frequency input of the supply network ($s = jw$).

Considering the former model already presented in Figure 3.13, that is equivalent to the following model, illustrated in Figure 3.18, when the below relations are considered:

$$L_{\sigma T1} = X_{\sigma T1}/w \qquad\qquad [3.22]$$

$$L'_{\sigma T2} = X'_{\sigma T2}/w \qquad\qquad [3.23]$$

$$L'_{\sigma D1} = X'_{\sigma D1}/w \qquad\qquad [3.24]$$

$$L_{hT} = X_{hT}/w \qquad\qquad [3.25]$$

$$L'_{hD} = X'_{hD}/w \qquad\qquad [3.26]$$

$$L''_{\sigma D2} = X''_{\sigma D2}/w \qquad\qquad [3.27]$$

Considering the resolver model – equations [3.19] to [3.21] (see Figure 3.17) – the system transfer function ($U''_2(s)/U_0(s)$) can be directly obtained by using the usual block diagram algebra:

$$\frac{U_2''(s)}{U_0(s)} = \frac{Z_3(s)}{Z_1(s) + Z_3(s)} \qquad\qquad [3.28]$$

where Z_1 and Z_3 have already been defined in the former equations [3.19] and [3.21]. Simplifying these impedances now (beginning by the variable Z_{Den}), we obtain, considering the notation from equations [3.22] to [3.27]:

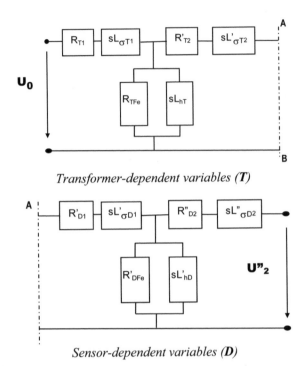

*Transformer-dependent variables (**T**)*

*Sensor-dependent variables (**D**)*

Figure 3.18. *Complete pancake resolver model*

$$Z_{\text{Den}} = \frac{\left(R'_{\text{T2}} + sL'_{\sigma\text{T2}} + R'_{\text{D1}} + sL'_{\sigma\text{D1}}\right) \times \left(\left(R_{\text{TFe}} + sL_{h\text{T}}\right) \times \left(R'_{\text{DFe}} + sL'_{h\text{D}}\right)\right)}{\left(\left(R_{\text{TFe}} + sL_{h\text{T}}\right) \times \left(R'_{\text{DFe}} + sL'_{h\text{D}}\right)\right)}$$

$$+ \frac{\left(\left(R_{\text{TFe}} + sL_{h\text{T}}\right) \times \left(R'_{\text{DFe}} + sL'_{h\text{D}}\right)\right) + \left(\left(R'_{\text{DFe}} + sL'_{h\text{D}}\right) \times \left(R_{\text{TFe}} + sL_{h\text{T}}\right)\right)}{\left(\left(R_{\text{TFe}} + sL_{h\text{T}}\right) \times \left(R'_{\text{DFe}} + sL'_{h\text{D}}\right)\right)} \qquad [3.29]$$

First simplifying the numerator of Z_{Den} (numZ_{Den}) we obtain:

$$\text{num}Z_{\text{Den}} = \left[\left(R'_{\text{T2}} + R'_{\text{D1}}\right) + \left(L'_{\sigma\text{T2}} + L'_{\sigma\text{D1}}\right)\right] s \times \left(R_{\text{TFe}} + sL_{h\text{T}}\right) \times \left(R'_{\text{DFe}} + sL'_{h\text{D}}\right)$$

$$+ \left(R'_{\text{DFe}} \times R_{\text{TFe}} \times sL_{h\text{T}}\right) + \left(R_{\text{TFe}} \times sL_{h\text{T}} \times sL'_{h\text{D}}\right)$$

$$+ \left(R_{\text{TFe}} \times R'_{\text{DFe}} \times sL'_{h\text{D}}\right) + \left(R'_{\text{DFe}} \times sL_{h\text{T}} \times sL'_{h\text{D}}\right) \qquad [3.30]$$

After several manipulations, numZ$_{Den}$ can be written as:

$$\text{numZ}_{Den} = \left[L'_{hD} \times L_{hT} \times \left(L'_{\sigma T2} + L'_{\sigma D1} \right) \right] s^3$$

$$+ \left\{ L'_{hD} \left[L_{hT} \left(R'_{T2} + R'_{D1} \right) + \left(L'_{\sigma T2} + L'_{\sigma D1} \right) R_{TFe} \right] \right.$$

$$\left. + R'_{DFe} L_{hT} \left(L'_{\sigma T2} + L'_{\sigma D1} \right) + R'_{DFe} L_{hT} L'_{hD} + R_{TFe} L_{hT} L'_{hD} \right\} s^2$$

$$+ \left\{ L'_{hD} R_{TFe} \left(R'_{T2} + R'_{D1} \right) + R'_{DFe} \left[\left(R'_{T2} + R'_{D1} \right) L_{hT} + \left(L'_{\sigma T2} + L'_{\sigma D1} \right) R_{TFe} \right] \right.$$

$$\left. + R'_{DFe} R_{TFe} L_{hT} + R'_{DFe} R_{TFe} L'_{hD} \right\} s + \left\{ R'_{DFe} R_{TFe} \left(R'_{T2} + R'_{D1} \right) \right\} \qquad [3.31]$$

Rewriting the numZ$_{Den}$ in a simplified way, we obtain:

$$\text{numZ}_{Den} = as^3 + bs^2 + cs + d \qquad [3.32]$$

where:

$$a = L'_{hD} \times L_{hT} \times \left(L'_{\sigma T2} + L'_{\sigma D1} \right)$$

$$b = \left\{ L'_{hD} \left[L_{hT} \left(R'_{T2} + R'_{D1} \right) + \left(L'_{\sigma T2} + L'_{\sigma D1} \right) R_{TFe} \right] \right.$$

$$\left. + R'_{DFe} L_{hT} \left(L'_{\sigma T2} + L'_{\sigma D1} \right) + R'_{DFe} L_{hT} L'_{hD} + R_{TFe} L_{hT} L'_{hD} \right\}$$

$$c = \left\{ L'_{hD} R_{TFe} \left(R'_{T2} + R'_{D1} \right) + R'_{DFe} \left[\left(R'_{T2} + R'_{D1} \right) L_{hT} + \left(L'_{\sigma T2} + L'_{\sigma D1} \right) R_{TFe} \right] \right.$$

$$\left. + R'_{DFe} R_{TFe} L_{hT} + R'_{DFe} R_{TFe} L'_{hD} \right\}$$

$$d = \left\{ R'_{DFe} R_{TFe} \left(R'_{T2} + R'_{D1} \right) \right\}$$

Finally, Z$_{Den}$ can be written in a simplified way:

$$Z_{Den} = \frac{as^3 + bs^2 + cs + d}{\left(\left(R_{TFe} + sL_{hT} \right) \times \left(R'_{DFe} + sL'_{hD} \right) \right)} \qquad [3.33]$$

Following the same methodology we can now simplify Z_1:

$$Z_1 = R_{T1} + sL_{\sigma T1} + \left[\frac{R_{TFe}L_{hT}(R'_{T2}+R'_{D1})s(R'_{DFe}+sL'_{hD})}{as^3+bs^2+cs+d} \right.$$

$$\left. +\frac{R_{TFe}L_{hT}(L'_{\sigma T2}+L'_{\sigma D1})s^2(R'_{DFe}+sL'_{hD})}{as^3+bs^2+cs+d} \right] \qquad [3.34]$$

First simplifying the Z_1 numerator (numZ_1) we obtain:

$$\text{num}Z_1 = [aL_{\sigma T1}]s^4 + [bL_{\sigma T1}+aR_{T1}+L'_{hD}(L'_{\sigma T2}+L'_{\sigma D1})R_{TFe}L_{hT}]s^3$$

$$+\left[cL_{\sigma T1}+bR_{T1}+L'_{hD}(R'_{T2}+R'_{D1})R_{TFe}L_{hT}\right.$$

$$\left.+R'_{DFe}(L'_{\sigma T2}+L'_{\sigma D1})R_{TFe}L_{hT}\right]s^2 +$$

$$+\left[dL_{\sigma T1}+cR_{T1}+R'_{DFe}(R'_{T2}+R'_{D1})R_{TFe}L_{hT}\right]s+dR_{T1} \qquad [3.35]$$

Finally, Z_1 can be written in a simplified way:

$$Z_1 = \frac{\text{num}Z_1}{as^3+bs^2+cs+d} \qquad [3.36]$$

Simplifying the last impedance Z_3 now we obtain:

$$Z_3 = \frac{(R_{TFe}L_{hT}R'_{DFe}L'_{hD})s^2}{as^3+bs^2+cs+d} \qquad [3.37]$$

Knowing that the relationship between the resolver output voltage U_{cos} and the voltage U'' is dependent on the device's global transformer ratio leads to the equation:

$$\frac{U_{cos}(s)}{U''(s)} = \frac{1}{r_T+r_D} \qquad [3.38]$$

where:

r_T = transformer ratio from transformer (T);

r_D = transformer ratio from sensor (D).

According to this equation we can finally derive the system transfer function $(U_{\cos}(s)/U_0(s))$:

$$\frac{U_{\cos}(s)}{U_0(s)} = \frac{1}{r_T r_D} \times \frac{Z_3(s)}{Z_1(s)+Z_3(s)} = \frac{1}{r_T r_D} \times \frac{\text{num}Z_3(s)}{\text{num}Z_1(s)+\text{num}Z_3(s)} \qquad [3.39]$$

$$\frac{U_{\cos}(s)}{U_0(s)} = \frac{1}{r_T r_D} \times \frac{\text{num}Z_3(s)}{F(s)} \qquad [3.40]$$

where:

$$\text{num}Z_3 = \left(R_{\text{TFe}} L_{hT} R'_{\text{DFe}} L'_{hT} S^2 \right)$$

$$F(s) = L_{\sigma T1} a S^4 + \left[L_{\sigma T1} b + R_{T1} a + L'_{hD} (L'_{\sigma T2} + L'_{\sigma D1}) R_{\text{TFe}} L_{hT} \right] S^3$$

$$+ \left[L_{\sigma T1} c + R_{T1} b + L'_{hD} R_{\text{TFe}} L_{hT} (R'_{T2} + R'_{D1}) \right.$$

$$\left. + R'_{\text{DFe}} R_{\text{TFe}} L_{hT} (L'_{\sigma T2} + L'_{\sigma D1}) \right] S^2 + \left[L_{\sigma T1} d + R_{T1} c \right.$$

$$\left. + R'_{\text{DFe}} R_{\text{TFe}} L_{hT} (R'_{T2} + R'_{D1}) \right] S + R_{T1} d$$

$$a = L'_{hD} L_{hT} (L'_{\sigma T2} + L'_{\sigma D1})$$

$$b = L'_{hD} \left[L_{hT} (R'_{T2} + R'_{D1}) + R_{\text{TFe}} (L'_{\sigma T2} + L'_{\sigma D1}) \right]$$

$$+ R'_{\text{DFe}} L_{hT} (L'_{\sigma T2} + L'_{\sigma D1}) + R_{\text{TFe}} L_{hT} L'_{hD} + R'_{\text{DFe}} L_{hT} L'_{hD}$$

$$c = L'_{hD} R_{\text{TFe}} (R'_{T2} + R'_{D1}) + R'_{\text{DFe}} R_{\text{TFe}} L_{hT} + R'_{\text{DFe}} R_{\text{TFe}} L'_{hD}$$

$$+ R'_{\text{DFe}} \left[L_{hT} (R'_{T2} + R'_{D1}) + R_{\text{TFe}} (L'_{\sigma T2} + L'_{\sigma D1}) \right]$$

$$d = R'_{\text{DFe}} R_{\text{TFe}} (R'_{T2} + R'_{D1})$$

The other important explicit model that will be derived here is the transfer function which relates the resolver current consumption (I) with its input voltage (U_0).

Considering the resolver model presented in Figure 3.17 (equations [3.19] – [3.21]), we can directly obtain the system transfer function (I(s)/U_0(s)) using the usual block diagram algebra:

$$\frac{I(s)}{U_0(s)} = \frac{1}{Z_1(s) + Z_3(s)} = \frac{\mathrm{den}Z_1(s)}{F(s)} \qquad [3.41]$$

where:

$$\mathrm{den}Z_1 = as^3 + bs^2 + cs + d$$

$$F(s) = L_{\sigma T1}aS^4 + \left[L_{\sigma T1}b + R_{T1}a + L'_{hD}\left(L'_{\sigma T2} + L'_{\sigma D1}\right)R_{TFe}L_{hT} \right]S^3$$

$$+ \left[L_{\sigma T1}c + R_{T1}b + L'_{hD}\,R_{TFe}L_{hT}\left(R'_{T2} + R'_{D1}\right) \right.$$

$$+ R'_{DFe}\,R_{TFe}L_{hT}\left(L'_{\sigma T2} + L'_{\sigma D1}\right)\bigg]S^2 + \left[L_{\sigma T1}d + R_{T1}c \right.$$

$$+ R'_{DFe}\,R_{TFe}L_{hT}\left(R'_{T2} + R'_{D1}\right)\bigg]S + R_{T1}d$$

$$a = L'_{hD}\,L_{hT}\left(L'_{\sigma T2} + L'_{\sigma D1}\right)$$

$$b = L'_{hD}\left[L_{hT}\left(R'_{T2} + R'_{D1}\right) + R_{TFe}\left(L'_{\sigma T2} + L'_{\sigma D1}\right) \right]$$

$$+ R'_{DFe}\,L_{hT}\left(L'_{\sigma T2} + L'_{\sigma D1}\right) + R_{TFe}L_{hT}L'_{hD} + R'_{DFe}\,L_{hT}L'_{hD}$$

$$c = L'_{hD}\,R_{TFe}\left(R'_{T2} + R'_{D1}\right) + R'_{DFe}\,R_{TFe}L_{hT} + R'_{DFe}\,R_{TFe}L'_{hD}$$

$$+ R'_{DFe}\left[L_{hT}\left(R'_{T2} + R'_{D1}\right) + R_{TFe}\left(L'_{\sigma T2} + L'_{\sigma D1}\right) \right]$$

$$d = R'_{DFe}\,R_{TFe}\left(R'_{T2} + R'_{D1}\right)$$

This obtained explicit mathematical model is suitable to test the accuracy of the assumed approach. In fact, all the parameters needed can be evaluated experimentally as will be explained in section 3.2.2.3.

The above model will be taken to compute the resolver nominal design variables (the standards for all product variables – $f(x_0)$).

In the next section we will introduce a new differential model to compute the influences on the main functional characteristics of a resolver – output voltage (U_{cos}, U_{sin}) and input current (I) – caused by small changes in product parameters, due to the variability of the assembly processes.

3.2.2.2. Incremental model – [$(\partial f / \partial x_i)_0 (\Delta x_i)$]

Having a general function f in R^n [$f(x_1, x_2, ..., x_n)$] this function can be linearized around the point ($x_{10}, x_{20}, ..., x_{n0}$) by cutting its Taylor's serial development after the first order partial derivatives, which leads to:

$$f\left(x_1, x_2, ..., x_n\right) = f\left(x_{10}, x_{20}, ..., x_{n0}\right) + \left.\frac{\partial f}{\partial x_1}\right|_0 \left(x_1 - x_{10}\right) + ... + \left.\frac{\partial f}{\partial x_n}\right|_0 \left(x_n - x_{n0}\right) \quad [3.42]$$

This methodology was used to build up a new mathematical model that was able to compute the influences on the resolver main functional characteristics: output voltage (U_{cos}, U_{sin}) and input current (I) caused by small changes on the manufacturer controllable variables (winding parameters).

The linear model presented here develops a complete new approach to model the product characteristics of a pancake resolver from the knowledge of the manufacturer controllable variables (winding parameters).

In Figure 3.19 the pancake resolver controllable model for a standard manufacturer is shown.

The variables considered by the are:

U_0 = resolver input voltage;

F = input frequency;

n_{st} = number of windings of the stator transformer;

n_{rt} = number of windings of the rotor transformer;

n_{ss} = number of windings of the stator sensor;

n_{rs} = number of windings of the rotor sensor;

ϕ_{st} = winding wire diameter of the stator transformer;

ϕ_{rt} = winding wire diameter of the rotor transformer;

ϕ_{ss} = winding wire diameter of the stator sensor;

ϕ_{rs} = winding wire diameter of the rotor sensor.

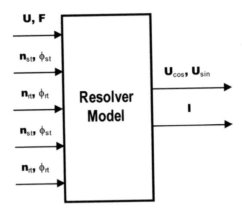

Figure 3.19. *Resolver controllable model*

The differential model for the resolver output voltage – U_{cos} – considers the marginal changes on the manufacturer controllable variables (winding parameters) and it can be computed as it is shown in equation [3.43].

Using the same approach, the influences on the input current – I – caused by small changes on the controllable variables (winding parameters) can be calculated as is illustrated in equation [3.44].

The variables with the subscript 0 refer to the nominal conditions computed by the nominal model, previously described in section – Model for nominal conditions [$f(x_0)$].

$$
U_{cos}\left(U, F, n_{st}, n_{rt}, n_{ss}, n_{rs}, \varphi_{st}, \varphi_{rt}, \varphi_{ss}, \varphi_{rs}\right)
$$

$$
= U_{cos}\left(U_0, F_0, n_{st0}, n_{rt0}, n_{ss0}, n_{rs0}, \varphi_{st0}, \varphi_{rt0}, \varphi_{ss0}, \varphi_{rs0}\right)\Big|_0
$$

$$
+ \frac{\partial U_{cos}}{\partial U}\Big|_0 (U - U_0) + \frac{\partial U_{cos}}{\partial F}\Big|_0 (F - F_0) + \frac{\partial U_{cos}}{\partial n_{st}}\Big|_0 (n_{st} - n_{st0}) + \frac{\partial U_{cos}}{\partial n_{rt}}\Big|_0 (n_{rt} - n_{rt0})
$$

$$
+ \frac{\partial U_{cos}}{\partial n_{ss}}\Big|_0 (n_{ss} - n_{ss0}) + \frac{\partial U_{cos}}{\partial n_{rs}}\Big|_0 (n_{rs} - n_{rs0}) + \frac{\partial U_{cos}}{\partial \phi_{st}}\Big|_0 (\phi_{st} - \phi_{st0}) + \frac{\partial U_{cos}}{\partial \phi_{rt}}\Big|_0 (\phi_{rt} - \phi_{rt0}) +
$$

$$
+ \frac{\partial U_{cos}}{\partial \varphi_{ss}}\Big|_0 (\varphi_{ss} - \varphi_{ss0}) + \frac{\partial U_{cos}}{\partial \varphi_{rs}}\Big|_0 (\varphi_{rs} - \varphi_{rs0}) \tag{3.43}
$$

$$I\left(U, F, n_{\mathrm{st}}, n_{\mathrm{rt}}, n_{\mathrm{ss}}, n_{\mathrm{rs}}, \varphi_{\mathrm{st}}, \varphi_{\mathrm{rt}}, \varphi_{\mathrm{ss}}, \varphi_{\mathrm{rs}}\right)$$

$$= I\left(U_0, F_0, n_{\mathrm{st0}}, n_{\mathrm{rt0}}, n_{\mathrm{ss0}}, n_{\mathrm{rs0}}, \varphi_{\mathrm{st0}}, \varphi_{\mathrm{rt0}}, \varphi_{\mathrm{ss0}}, \varphi_{\mathrm{rs0}}\right)\Big|_0$$

$$+ \frac{\partial I}{\partial U}\Big|_0 (U - U_0) + \frac{\partial I}{\partial F}\Big|_0 (F - F_0) + \frac{\partial I}{\partial n_{\mathrm{st}}}\Big|_0 (n_{\mathrm{st}} - n_{\mathrm{st0}}) + \frac{\partial I}{\partial n_{\mathrm{rt}}}\Big|_0 (n_{\mathrm{rt}} - n_{\mathrm{rt0}})$$

$$+ \frac{\partial I}{\partial n_{\mathrm{ss}}}\Big|_0 (n_{\mathrm{ss}} - n_{\mathrm{ss0}}) + \frac{\partial I}{\partial n_{\mathrm{rs}}}\Big|_0 (n_{\mathrm{rs}} - n_{\mathrm{rs0}}) + \frac{\partial I}{\partial \phi_{\mathrm{st}}}\Big|_0 (\phi_{\mathrm{st}} - \phi_{\mathrm{st0}}) + \frac{\partial I}{\partial \phi_{\mathrm{rt}}}\Big|_0 (\phi_{\mathrm{rt}} - \phi_{\mathrm{rt0}})$$

$$+ \frac{\partial I}{\partial \varphi_{\mathrm{ss}}}\Big|_0 (\varphi_{\mathrm{ss}} - \varphi_{\mathrm{ss0}}) + \frac{\partial I}{\partial \varphi_{\mathrm{rs}}}\Big|_0 (\varphi_{\mathrm{rs}} - \varphi_{\mathrm{rs0}}) \qquad [3.44]$$

The several partial derivatives presented in both equations [3.43] and [3.44] have been experimentally identified, with a set of measuring points, which were fitted by second order polynomials (see next section).

3.2.2.3. Parameter identification

The parameter identification for the model developed in the previous section was performed experimentally at Tyco Electronics – Évora plant, applied to the standard pancake 1–speed resolver, with reference H2109, where nominal specifications are:

Input voltage $(U_0) = 5$ V;

Nominal frequency = 4 kHz;

Max. input current = 50 mA.

3.2.2.3.1. Nominal model

In order to evaluate the nominal model parameters we tested this resolver in both cases: open and short circuited. In the open-circuit case the impedances R and X_σ are much smaller than the parallel impedance $R_{\mathrm{Fe}}//X_h$, then the complete model from Figure 3.12 reduces to the model presented in Figure 3.20.

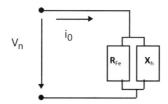

Figure 3.20. *Reduced transformer model (open circuit)*

In the short-circuit case the model from Figure 3.12 reduces to the model presented in Figure 3.21. These computing approaches were applied to both pancake resolver main components: transformer and sensor.

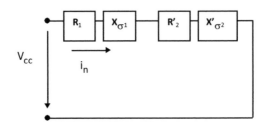

Figure 3.21. *Reduced transformer model (short circuit)*

The values measured for both transformer and sensor are presented in Table 3.1.

Transformer			
Open circuit		Short circuit	
V_n [V]	5.00		
V_2 [V]	6.72	V_{cc} [V]	3.90
I_0 [mA]	14.91	I_n [mA]	50.00
P_0 [mW]	40.00	P_{cc} [mW]	73.00
$R_1 = 20.2$ Ohm			
$R_2 = 19.0$ Ohm			
Sensor			
Open circuit		Short circuit	
V_n [V]	3.70		
V_2 [V]	2.84	V_{cc} [V]	1.49
I_0 [mA]	14.51	I_n [mA]	20.30
P_0 [mW]	20.00	P_{cc} [mW]	18.00
$R_1 = 19.0$ Ohm			
$R_2 = 21.0$ Ohm			

Table 3.1. *Experimental values*

Using these experimental values together with the standard circuit analysis theory, on both the above models (from Figures 3.20 and 3.21), we identify finally the resolver nominal model parameters, presented in Table 3.2. These computed parameters refer to equations [3.40] and [3.41].

Transformer		Sensor	
R_{T1} [Ohm]	19.21	R_{D1} [Ohm]	8.44
$X_{\sigma T1}$ [Ohm]	47.57	$X_{\sigma D1}$ [Ohm]	11.40
R_{TFe} [Ohm]	625.00	R_{DFe} [Ohm]	379.27
X_{hT} [Ohm]	397.46	X_{hD} [Ohm]	152.04
R'_{T2} [Ohm]	9.99	R'_{D2} [Ohm]	15.75
$X_{\sigma'T2}$ [Ohm]	24.73	$X_{\sigma'D2}$ [Ohm]	21.26

Table 3.2. *Pancake resolver nominal model parameters*

3.2.2.3.2. Incremental model

In order to evaluate the incremental model parameters we need to know all the several partial derivatives calculated at the nominal conditions. Here we face a problem because we do not have these mathematical functions. To solve this problem we identified these partial derivatives by experimental measurements and then numerical approximations by second order polynomials $[a + b(x-x_m) + c(x-x_m)^2]$ were accomplished.

Figure 3.22 shows an example of the previously described methodology applied to the identification of the partial derivative $-\partial I/\partial n_{st}-$ (input current elasticity in relation to the number of windings of stator transformer) with all other variables set to the nominal values.

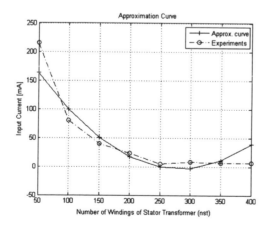

Figure 3.22. *Parameter identification in incremental model – function $\partial I/\partial n_{st}$*

Following the interpolation procedure, we obtain the function:

$$\frac{\partial I}{\partial n_{st}} = 10.41 - 0.42(n_{st} - 215) + 0,0031(n_{st} - 215)^2 \qquad [3.45]$$

The procedure described above was applied to identify all the partial derivatives presented in equations [3.43] and [3.44].

3.3. Simulation and experimental results

The results presented in this chapter are split into two groups:

– performance of the overall model (nominal + incremental);

– manufacturer correction tools (incremental model).

3.3.1. *Performance of the overall model*

The results delivered by the overall model are presented in Figures 3.23 and 3.24. All computations have been done using the Matlab software [MAT].

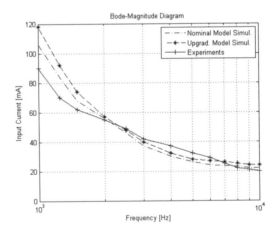

Figure 3.23. *Frequency response results from resolver's input current magnitude*

Here we can see the experimental measurements and the simulated values delivered by each model separately – nominal and upgraded models – (upgraded model = nominal model + incremental model).

As was expected, the main contribution is due to the nominal model and the incremental model shows a comparatively reduced contribution although it approximates the simulated values to the measured points.

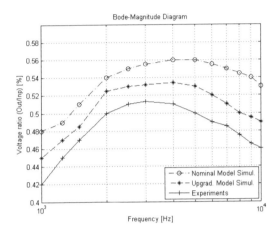

Figure 3.24. *Frequency response results from resolver's voltage ratio magnitude*

3.3.2. *Manufacturer correction tools*

The results that are shown in this section refer to the incremental model and they were nominated as manufacturer correction tools, as these results are effectively the new developed tools that are available to the resolver manufacturer in order to correct the deviations on product characteristics caused by changes in the production factors (manufacturing processes, materials, etc.).

In fact, a set of correction tools dependent on manufacturer-controlled variables (winding parameters) was developed, and can be used to compensate for deviations in product characteristics.

The knowledge of the developed correction tools permits the resolver manufacturer to change some controllable variables (usually the number of windings in transformers) in order to correct assembled resolvers that without any action would be scrap to the production line (usually input current or output voltage out of specifications).

Figures 3.25–3.30 present the simulated and experimental results from some of the developed functions that show the resolver manufacturer the way to act on the correspondent controllable variable in order to influence the resolver main customer characteristics (input current – I, and output voltage – $U_{\cos}$).

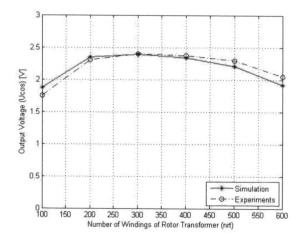

Figure 3.25. *Output voltage ($U_{cos(0)}$) vs. number of windings of the rotor transformer (n_{rt})*

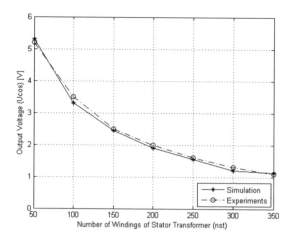

Figure 3.26. *Output voltage ($U_{cos(0)}$) vs. number of windings of the stator transformer (n_{st})*

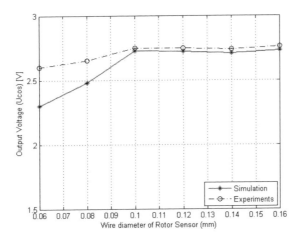

Figure 3.27. *Output voltage ($U_{cos(0)}$) vs. winding wire diameter of the rotor sensor (ϕ_r)*

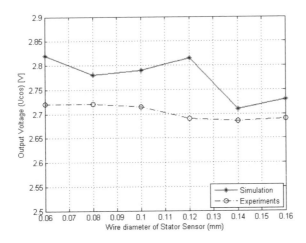

Figure 3.28. *Output voltage ($U_{cos(0)}$) vs. winding wire diameter of the stator sensor (ϕ_s)*

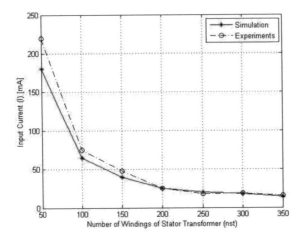

Figure 3.29. *Input current ($I_{(0)}$) vs. number of windings of the stator transformer (n_{st})*

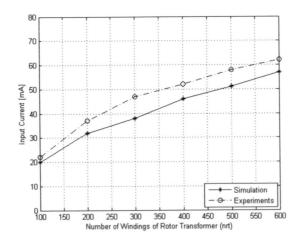

Figure 3.30. *Input current ($I_{(0)}$) vs. number of windings of the rotor transformer (n_{rt})*

3.4. Conclusions

The two-step strategy followed to build up the resolver mathematical model (traditional transformer's model upgraded with a linear controllable model) proved to be very efficient to the resolver manufacturer.

The newly developed linear model that relates the resolver main electrical characteristics (input current and output voltage) to the manufacturer-controllable variables – basically the winding parameters – allows the development of a set of correction tools that allow the resolver manufacturer to change some controllable variables in order to correct assembled resolvers that without any action would be scrap to the production line.

The denominated correction tools are partial derivatives of the resolver's main electrical characteristics (input current and output voltage). These partial derivatives reflect the sensitivity of the resolver's main electrical characteristics to each one of the production controllable variables. This knowledge allows the manufacturer to react quickly to product deviations due to unknown changes in the production processes.

3.5. Acknowledgment

All laboratorial work presented in this chapter was supported by means of manufacturing machinery, test equipment and resolver products by Tyco Electronics – Plant Évora. This work results from an existing cooperation program established between University of Evora and Tyco Electronics – Plant Évora.

The work was supported by Fundação para a Ciência e Tecnologia, through IDMEC under LAETA.

3.6. Bibliography

[ALH 04] ALHAMADI M., BENAMMAR M., BEN-BRAHIM L., "Precise method for linearizing sine and cosine signals in resolvers and quadrature encoders applications", *Proceedings of 30th Annual Conference of IEEE Industrial Electronics Society (IECON 2004)*, Busan, South Korea, 2–6 November 2004.

[ATT 07] ATTAIANESE C., TOMASSO G., "Position measurement in industrial drives by means of low-cost resolver-to-digital converter", *IEEE Transactions on Instrumentation and Measurement*, vol. 56, no. 6, December 2007, pp. 2155–2159.

[BEN 04] BENAMMAR M., BEN-BRAHIM L., ALHAMADI M., "A novel resolver-to-360 degree linearized converter", *IEEE Sensors Journal*, vol. 4, 2004, pp. 96–101.

[BEN 05] BENAMMAR M., BEN-BRAHIM L., ALHAMADI M., "A high precision resolver-to-DC converter", *IEEE Transactions on Industrial Electronics*, vol. 54, no. 6, December 2005, pp. 2289–2296.

[BEN 07] BENAMMAR M., BEN-BRAHIM L., ALHAMADI M., AL-NAEMI M., "A novel method for estimating the angle from analog co-sinusoidal quadrature signals", *Sensors and Actuators A: Physical*, March 2007, Available: http://www.sciencedirect.com.

[BER 03] BERTRAM T., BEKES F., GREUL R., HANKE O., HASS C., HILGERT J., HILLER M., OETTGEN O., REIN P., TORLO M., WARD D., "Modeling and simulation for mechatronic design in automotive systems", *Control Engineering Practice*, vol. 11, 2003, pp. 179–190.

[BRA 08] BEN-BRAHIM L., BENAMMAR M., ALHAMADI M., AL-EMADI N., "A new low cost linear resolver converter", *IEEE Sensors Journal*, vol. 8, no. 10, October 2008, pp. 1620–1627.

[BRA 09] BEN-BRAHIM L., BENAMMAR M., ALHAMADI M., "A resolver angle estimator based on its excitation signal", *IEEE Transactions on Industrial Electronics*, vol. 56, no. 2, February 2009, pp. 574–580.

[BUN 04] BUNTE A., BEINEKE S., "High-performance speed measurement by suppression of systematic resolver and encoder errors", *IEEE Transactions on Industrial Electronics*, vol. 51, no. 1, February 2004, pp. 49–53.

[BUR 08] BURROW S., MELLOR P., CHURN P., SAWATA T., HOLME M., "Sensorless operation of a permanent-magnet generator for aircraft", *IEEE Transactions on Industry Applications*, vol. 44, no. 1, January 2008, pp. 101–107.

[FIG 04] FIGUEIREDO J., SÁ DA COSTA J., "Analog rotation sensors: an industrial approach for modelling and simulation", *Proceedings of 6th International Conference on Automatic Control, IFAC/APCA*, Faro, Portugal, 7–9 June 2004.

[GOL 81] GOLKER W., TAMM H., SMITH S., "Ein messgetriebe fuer den airbus A310: position pick-off unit", *Siemens Components 19*, 1981, Heft 4, pp. 125–128.

[HAN 90] HANSELMAN D., "Resolver signal requirements for high accuracy resolver-to-digital conversion", *IEEE Transactions on Industrial Electronics*, vol. 37, no. 6, December 1990, pp. 556–561.

[HOS 07] HOSEINNEZHAD R., BAB-HADIASHAR A., HARDING P., "Calibration of resolver sensors in electromechanical braking systems: A modified recursive weighted least-squares approach", *IEEE Transactions on Industrial Electronics*, vol. 54, no. 2, April 2007, pp. 1052–1060.

[KIM 09] KIM K., SUNG C., LEE J., "Magnetic shield design between interior permanent magnet synchronous motor and sensor hybrid electric vehicle", *IEEE Transactions on Magnetics*, vol. 45, no. 6, 2009, pp. 2835–2838.

[MAP 10] MAPELLI F., TARSITANO D., MAURI M., "Plug-in hybrid electric vehicle: modeling, prototype realization, and inverter losses reduction analysis", *IEEE Transactions on Industrial Electronics*, vol. 57, no. 2, February 2010, pp. 598–607.

[MAS 00] MASAKI K., KITAZAWA K., MIMURA H., NIREI M., TSUCHIMICHI K., WAKIWAKA H., YAMADA H., "Magnetic field analysis of a resolver with a skewed and eccentric rotor", *Sensors and Actuators A: Physical*, March 2000, Available: http://www. sciencedirect.com.

[MAT] Mathworks, *Matlab*, The Math Works.

[MUR 02] MURRAY A., HARE B., HIRAO A., "Resolver position sensing system with integrated fault detection for automotive applications", *Proceedings of IEEE Sensors*, vol. 2, 2002, pp. 864–869.

[SAR 08] SARMA S., AGRAWAL V., UDUPA S., "Software-based resolver-to-digital conversion using a DSP", *IEEE Transactions on Industrial Electronics*, vol. 55, no. 1, January 2008, pp. 371–379.

[SIE 96] SIEMENS E., HALSKE J., Verfahren zur Fernuebertragung von Bewegungen, Patent application no. 93912, Germany, 1896.

[SUN 08] SUN L., "Analysis and improvement on the structure of variable reluctance resolvers", *IEEE Transactions on Magnetics*, vol. 44, no. 8, August 2008, pp. 2002–2008.

[VAZ 10] VAZQUEZ N., LOPEZ H., HERNANDEZ C., VAZQUEZ E., OSORIO R., ARAU J., "A different multilevel current-source inverter", *IEEE Transactions on Industrial Electronics*, vol. 57, no. 8, August 2010, pp. 2623–2632.

Chapter 4

Robust Control of Atomic Force Microscopy

The atomic force microscope (AFM) is an instrument used for acquiring images at nanometer scale. Obtaining better image quality at higher scan speed is a research area of great interest in the control of an AFM. Improving the dynamic response of the scanning probe in the vertical direction and the dynamic response of the scanning motion in the lateral plane are the two major areas of application of advanced control methods to an AFM. The uncertainties inherent in the models of AFM vertical and lateral direction motion stages dictates the application of robust control methods. In this chapter, robust control methods are applied to AFM, treating first the vertical direction and then the lateral plane.

4.1. Introduction

Improvement in the AFM stage dynamics is achieved either by designing stages with higher bandwidth or by designing more sophisticated controllers rather than in the PI, PID, or PIID types of controllers that are most commonly used in practice. A robust repetitive controller is used for the vertical direction as it can reject higher frequency disturbances due to the periodic part of the surface topography in AFM much better than a conventional controller. Besides increasing the scan speed, it is also important that the phase lag can be compensated using repetitive control, with the knowledge of the surface topography from the previous period by introducing appropriate phase advance into the controller. Next, a multi-input–multi-output (MIMO) extension of the disturbance observer control method is applied to the

Chapter written by Bilin AKSUN GÜVENÇ, Serkan NECİPOĞLU, Burak DEMİREL and Levent GÜVENÇ.

lateral plane of scanning motion in a piezoelectric tube-based atomic force microscope (AFM). Calibration free and decoupled operation of the AFM is achieved with this technique. The technique is also robust to creep and hysteresis effects that are common in piezoelectric actuators. Both the repetitive and the MIMO disturbance observer controllers are designed using the control of mechatronic systems toolbox (COMES).

4.2. Repetitive control of the vertical direction motion

The AFM invented by Binnig *et al.* [BIN 86] is used for acquiring surface topography at the precision of nanometers. The selective features of AFM such as the ability of fast and easy sample preparation, air, liquid, and vacuum environments of operation, relatively lower costs, and so on make it an imaging technique of strong preference. Hence, improving the performance of AFM scanning has been an active area of research. The performance of an AFM can be described in terms of its scanning speed and image quality, which are inversely proportional to each other. The two major limitations imposed on scanning speed without violating the image quality and stability are the transient response of the cantilever probe and the mechanical bandwidth of the mechanisms used on the vertical axis "z", which are mostly made of piezoelectric actuators. These are followed by the general limitations of the feedback loop such as time delays, sampling rate in the case of digital control, sensor noise, RMS conversion rate, and so on. More information on AFM dynamics and control is given in [GAR 02] and [ABR 07]. Figure 4.1 shows a basic presentation of an AFM setup.

The transient response of the probe is quantified by the quality factor (Q) of the cantilever beam [SUL 02]. High Q values cause slow response of the probe to surface topographic changes and even instability in dynamic, amplitude modulated (AM) AFM, e.g. tapping mode. An active Q control to improve the response time is also proposed in [SUL 02]. An adaptive Q control (AQC) depending on the surface properties is proposed in [GUN 07], whereas a full state feedback control method affecting both Q and the resonant frequency of the vibrating probe is presented in [ORU 09].

It is very common among physicists to use a PI, PID, PII, or a PIID controller for the vertical motion of the scanner on z-direction [ABR 07]. Obviously, a simple PI controller cannot improve the bandwidth to perform good surface tracking at high frequencies. Adding a derivative term seems to be a good idea at first, but this is avoided because the measurement of the probe's deflection is noisy. However, the bandwidth of the scanner's vertical motion in the z-direction can be improved by using a more sophisticated mechanical design [SCH 07] or by implementing more advanced control techniques. Such an advanced controller is implemented in

[SCH 01] utilizing H_∞ control theory. Advanced robust controllers can handle the inevitable nonlinearities and system uncertainties as well.

When the continuity of the scanned surface is considered, it is reasonable to assume that the successive lines of the scan are similar. This motivates researchers to make use of past scanning information for improving the performance of the scan on following scan lines. The combination of feedforward and H_∞ controller is used in [SCH 04] for this purpose. Other feedforward, learning- and observer-based controllers are proposed in [SCH 04], [LI 08], [FUJ 08], and [SAL 05] for the periodic motions of the scanner. A brief discussion about the combination of feedback and feedforward controllers is presented in [PAO 07].

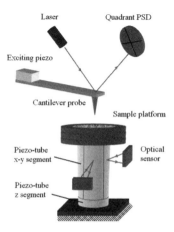

Figure 4.1. *Basic AFM setup: the probe is excited by a piezoelectric element using a sinusoidal wave form. The probe's deflection is measured by sensing the displacement of the laser beam reflected from the tip onto a photo sensor diode. The sample to be scanned is placed on a piezotube. The upper quarter part of the piezotube is used for the raster scan motion in the x–y plane and the lower single part is used for the vertical motion in z*

Having the same reasonable assumption made for the feedforward controllers, this paper focuses on the repetitive control technique which is a powerful way of tracking or rejecting periodic signals [AKS 06]. The organization of the rest of the section is as follows. In section 4.2.1, a tapping mode AFM system scheme is introduced along with a description of the experimental AFM hardware being used. Repetitive control basics and mapping the design specifications into parameter space are explained in sections 4.2.2 and 4.2.3, respectively. The repetitive control features of the COMES are outlined in section 4.2.4. In section 4.2.5, a parameter space-based robust repetitive controller is designed using the COMES toolbox running in Matlab. Simulation results obtained using an accurate and realistic computer model are demonstrated in section 4.2.6.

4.2.1. *Tapping mode AFM system model*

The model used in this part of the chapter on controlling the vertical axis of an AFM stage is based qualitatively on the numerical model in [VAR 08] that was built for simulating a tapping mode AFM. The complete system in [VAR 08] can be redrawn as in Figure 4.2 for control purposes. The scanning probe is vibrated at the frequency of 221 kHz. The excitation signal is adjusted to maintain a free air amplitude of 45 nm and the Q factor is set to 79. The reason for the selection of this Q will be explained later.

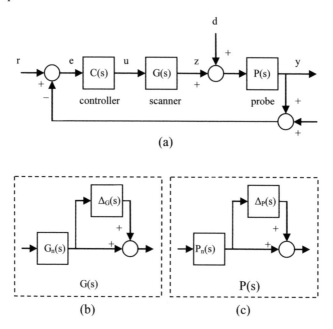

Figure 4.2. *(a) AFM system model in classical control where r is the set amplitude of the probe vibration, y is the actual amplitude, u is the control signal in volts, z is the scanner's vertical motion in nanometers, and d is the surface topography; (b) the uncertainty in the linear scanner model; and (c) the uncertainty in the linear probe model*

The interaction forces between the tip and the sample surface change the amplitude of vibration in tapping mode AFM. Since it is desired to keep these interaction forces unchanged, it is necessary to keep the amplitude of the probe's vibration unchanged. This is achieved by controlling the distance between the tip and the surface by feeding back this amplitude. This is also called constant force scanning. Thus, the probe works like a sensor of interaction forces such that the input is the distance, and the output is the amplitude of the vibration as shown in Figure 4.1.

However, the probe block in Figure 4.2a is not linear. The nonlinearities are due to the attractive and repulsive regimes of the interaction forces, the probe indentation into the sample, and so on [VAR 08]. The reader should note that all of these factors are present in the model that is used for simulations here. It is observed during numerical simulations that the probe shows a quite linear behavior for input signals greater that 10 nm with an approximate DC gain of unity. For those sizes of the inputs, the dynamics can be characterized by a first-order filter due to the sharp 90 degrees phase transition observed at around 3 kHz. The nonlinearities occur such that the DC gain converges to 2 and the dynamics become oscillatory in the order of 2 for closer proximity of the tip to the surface.

We do not want to be concerned with the probe's dynamics at this moment as our aim is to improve the bandwidth of the scanner's vertical axis motion, which typically lies within the 1–40 kHz frequency band. That is why the Q value is chosen to be low. Therefore, we neglect the uncertainties described in Figure 4.2c in the linear analysis and assume a static $P(s) = 1$ model for the probe dynamics since we will discuss the areas satisfying that condition. Note that robust handling of the probe nonlinearities can be achieved using the disturbance observer method presented in section 4.3.

Different from [VAR 08], the stage dynamics is chosen as that of a piezotube actuator here, as given in [OHA 95]. The transfer function of the piezotube's vertical axis motion is given in equation [4.1] where the input is the driving voltage and the output is the displacement along the z-axis in meters.

$$\frac{z_{pt}(s)}{V_z(s)} = \frac{158.7}{s^2 + 1328s + 1.763 \times 10^{10}} \tag{4.1}$$

The model in [OHA 95] is derived by curve fitting up to the first mode resonance frequency. The higher frequencies involve uncertainties as is most often the case when a high-order system is modeled using a reduced order representation.

The controller block is designed to keep the output of the feedback loop at the reference value, which is the set amplitude of the probe's vibration. The amplitude is calculated by RMS conversion after 10 oscillations of the probe. The amplitude sensor dynamics are usually of the order of MHz. So, it is safely taken as a static gain in the numerical model.

It is clear from Figure 4.2 that the surface topography is considered as a disturbance that should be rejected. The scanner's motion along the z-axis is recorded while performing the scan operation, thereby obtaining a record of the surface topography also. The complicated AFM system in [VAR 08] is viewed as a very common and well-known control problem with some assumptions made on the sensor probe and the piezotube actuator. These assumptions lead to simple linear models with uncertainty and the need to use a robust controller.

4.2.2. Repetitive control basics

The repetitive control structure is shown in Figure 4.3 where G_n is the nominal model of the plant, Δ_m is the normalized unstructured multiplicative model uncertainty, W_T is the multiplicative uncertainty weighting function, and τ_d is the period of the periodic exogenous signal. $q(s)$ and $b(s)$ are filters used for tuning the repetitive controller. Repetitive control systems can track periodic signals very accurately and can reject periodic disturbances very satisfactorily. This is due to the fact that the positive feedback loop in Figure 4.4 is a generator of periodic signals with period τ_d for $q(s) = 1$. A low-pass filter with unity DC gain is used for $q(s)$ for robustness of stability [HAR 88] and [WEI 97].

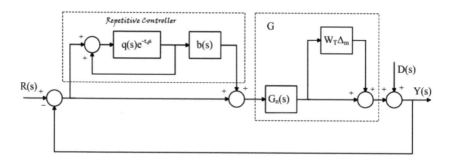

Figure 4.3. *Repetitive contolled system*

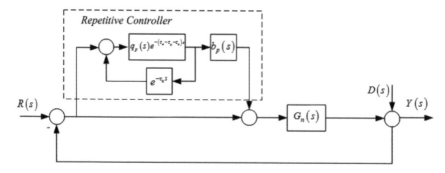

Figure 4.4. *Modified repetitive control system*

The repetitive controller design involves the design of the two filters $q(s)$ and $b(s)$ in Figure 4.3. In the frequency domain, the ideal low-pass filter $q(j\omega)$ would be unity in the frequency range of interest and zero at higher frequencies. This is not possible and $q(j\omega)$ will have negative phase angle which will make $q(j\omega)$ differ from unity, resulting in reduced accuracy. To improve the accuracy of the repetitive controller, a small time advance is customarily incorporated into $q(s)$ to cancel out the negative phase of its low-pass filter part within its bandwidth. This small time advance can easily be absorbed by the much larger time delay τ_d corresponding to the period of the exogenous input signal and does not constitute an implementation problem.

The main objective of the usage of the dynamic compensator $b(s)$ is to improve the relative stability, the transient response, and the steady-state accuracy in combination with the unity DC gain low-pass filter $q(s)$. Consider the function of frequency given by:

$$R(\omega) = \left| q(j\omega) \left[1 - b(j\omega) \frac{G(j\omega)}{1 + G(j\omega)} \right] \right| \qquad [4.2]$$

which is called the regeneration spectrum in [SRI 91]. According to the same reference, $R(\omega) < 1$ for all ω is a sufficient condition for stability. Moreover, $R(\omega)$ can be utilized to obtain a good approximation of the locus of the dominant characteristic roots of the repetitive control system for large time delay, thus resulting in a measure of relative stability, as well. Accordingly, the compensator $b(s)$ is designed to approximately invert $G/(1+G)$ within the bandwidth of $q(s)$ in an effort to minimize $R(\omega)$. The dynamic compensator $b(s)$ can be selected as only a small time advance or time advance multiplied by a low-pass filter to further minimize $R(\omega)$. To make $R(\omega) < 1$, the time advance in the filter $b(s)$ is chosen to cancel out the negative phase of $G/(1+G)$. This small time advance can easily be absorbed by the much larger time delay τ_d corresponding to the period of the exogenous input signal and does not constitute an implementation problem (Figure 4.4).

The $q(s)$ and $b(s)$ filters are thus expressed as:

$$q(s) = q_p(s)e^{\tau_q s} \qquad [4.3]$$

$$b(s) = b_p(s)e^{\tau_b s} \qquad [4.4]$$

The time advances, τ_q and τ_b, are first chosen to decrease the magnitude of $R(\omega)$ given in equation [4.2]. Then, the design focuses on pairs of chosen parameters in $q_p(s)$ or $b_p(s)$ to satisfy a frequency domain bound on the mixed sensitivity performance criterion. If L denotes the loop gain of a control system, its sensitivity and complementary sensitivity transfer functions are:

$$S = \frac{1}{1+L} \qquad\qquad [4.5]$$

$$T = \frac{L}{1+L} \qquad\qquad [4.6]$$

The parameter space design, presented in the following aims at satisfying the mixed sensitivity performance requirement:

$$\left\| |W_S S| + |W_T T| \right\|_\infty < 1 \quad \text{or} \quad |W_S S| + |W_T T| < 1 \quad \text{for } \forall \omega \qquad [4.7]$$

where W_S and W_T are the sensitivity and complementary sensitivity function weights. The loop gain of the repetitive control system seen in Figures 4.3 and 4.4 are given by:

$$L = G\left(1 + \frac{q_p}{1 - q_p e^{(-\tau_d + \tau_q)s}} b_p e^{(-\tau_d + \tau_q + \tau_b)s} \right) \qquad [4.8]$$

The mixed sensitivity design requires:

$$|W_S(\omega)S(j\omega)| + |W_T(\omega)T(j\omega)| = \left| \frac{W_S(\omega)}{1+L(j\omega)} \right| + \left| \frac{W_T(\omega)L(j\omega)}{1+L(j\omega)} \right| < 1 \qquad [4.9]$$

or equivalently equation [4.10] to be satisfied for all ω.

$$|W_S(\omega)| + |W_T(\omega)L(j\omega)| < |1 + L(j\omega)| \qquad [4.10]$$

4.2.3. Mapping mixed sensitivity specifications into controller parameter space

A repetitive controller design procedure based on mapping the mixed sensitivity frequency domain performance specification given in equation [4.10] with an equality sign into the chosen repetitive controller parameter plane at a chosen frequency is described here. Consider the mixed sensitivity problem given in Figure 4.5 illustrating equation [4.10] with an equality sign (called the mixed sensitivity point condition). Apply the cosine rule to the triangle with vertices at the origin, -1 and L in Figure 4.5, to obtain:

$$\left(|W_S(\omega)| + |W_T(\omega)L(j\omega)| \right)^2 = |L(j\omega)|^2 + 1^2 + 2|L(j\omega)|\cos\theta_L \qquad [4.11]$$

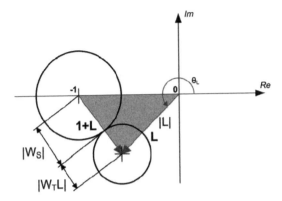

Figure 4.5. *Illustration of the point condition for the mixed sensitivity*

Equation [4.11] is quadratic in $|L(j\omega)|$ and its solutions are:

$$|L(j\omega)| = \frac{(-\cos\theta_L + |W_S(\omega)||W_T(\omega)|) \pm \sqrt{\Delta_M(\omega)}}{1 - |W_T(\omega)|^2} \qquad [4.12]$$

where:

$$\Delta_M(\omega) = \cos^2\theta_L + |W_S(\omega)|^2 + |W_T(\omega)|^2 - 2|W_S(\omega)||W_T(\omega)|\cos\theta_L - 1 \qquad [4.13]$$

Only positive and real solutions for $|L|$ are allowed, that is, $\Delta_M \geq 0$ in equation [4.12] must be satisfied. A detailed explanation of the point condition solution is given in [DEM 10].

4.2.4. *Repetitive control features of COMES*

COMES toolbox is a graphical user interface (GUI) for the routines of four different control approaches [DEM 09]: classical control (lead, lag, PID, and so on), preview control, model regulator control, and repetitive control, which are coded as Matlab M-files. The repetitive control design module of the COMES toolbox is used for determining the parameter space regions corresponding to chosen frequency-domain criteria. The solution technique is based on mapping a frequency domain mixed sensitivity bound into the chosen repetitive controller parameter plane as explained in the previous section. The procedure leads to graphical solution regions in 2D plots for each design specification. A screenshot from the repetitive control design module of COMES is shown in Figure 4.6.

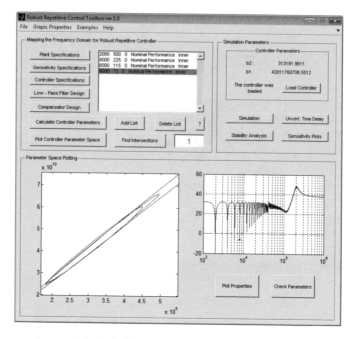

Figure 4.6. *GUI of the repetitive control module of COMES*

First, the plant specifications are introduced. Second, the sensitivity and complementary sensitivity weights are introduced for a range of discrete frequency values. Then, the controller specifications such as the fundamental or the harmonics of the repetitive signal and the number of grid points for the θ sweep in Figure 4.5 are entered. The q filter, which is a second-order low-pass filter with unity DC gain as described previously, is entered parametrically in terms of a_{00} and a_{01}. Then the compensator b is entered as presented previously as well. Finally, the low-pass filter parameters are calculated numerically by the COMES and the solution region satisfying the design criteria is plotted in the parameter space. Having repeated this calculation by updating the sensitivity specifications and the controller specifications for each frequency, new solution regions are plotted on the same plane. The overall solution region satisfying all the design criteria is the intersection of these regions, which is shown with color filling. The q filter parameters are then chosen by the user within the solution region.

The frequency plots of sensitivity, complementary sensitivity, loop gain, and the regeneration spectrum can be observed for convenience using the "Sensitivity Plots" pane. The aim of COMES is to provide a user-friendly toolbox with an interactive GUI that lets all necessary calculations run smoothly in the background while the user can focus on analyzing the graphical results.

4.2.5. *Robust repetitive controller design using the COMES toolbox*

The point condition solution is implemented using the repetitive control module of the COMES toolbox. The design specifications are determined as in Figure 4.7, as good tracking (nominal performance) at low frequencies, mixed sensitivity at intermediate frequencies, and robust stability at high frequencies where unstructured multiplicative uncertainties of the piezotube exist. No performance specification is required near the resonance since the system is not operated at those frequencies.

The weights W_S and W_T are determined for arbitrary frequencies inside the regions in Figure 4.7 as shown in Table 4.1. The design is based on a periodic signal with a fundamental frequency of 200 Hz, hence the repeating period is 0.005 s.

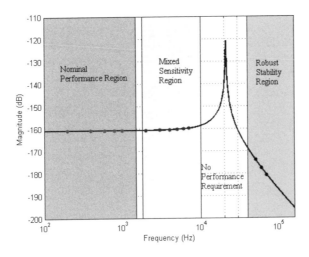

Figure 4.7. *Design specifications for the repetitive controller*

Having determined the design specifications in Table 4.1, the filter $b(s)$ is designed as in [4.14], and $q(s)$ is chosen to be in the form of [4.15]. The parameters a_{00} and a_{01} in [4.15] must be appropriately selected from the parameter space to satisfy the design specifications given in Figure 4.7 and Table 4.1. The solution regions of the point condition on the parameter space are plotted in Figure 4.8 for each frequency given in Table 4.1. The intersection of those regions is color filled and an arbitrary point somewhere near the center of this intersection is selected to determine a_{00} and a_{01} that are given in [4.16].

$$b(s) = \frac{6.707 \times 10^7 s^2 + 1.591 \times 10^{14} s + 1.104 \times 10^{18}}{s^2 + 2 \times 10^5 s + 10^{10}} \qquad [4.14]$$

$$q(s) = \frac{a_{00}}{s^2 + a_{01}s + a_{00}} \qquad\qquad [4.15]$$

$$a_{00} = 3.8882 \times 10^8; \quad a_{01} = 2.9133 \times 10^4 \qquad\qquad [4.16]$$

Finally, time advances $\tau_b = 6.268 \times 10^{-6}$s and $\tau_q = 7.5 \times 10^{-5}$s are calculated to compensate for the phase lags introduced by $q(s)$ and $b(s)G(s)/[1+G(s)]$ as shown in Figure 4.9.

$f = k/\tau_d$ (Hz)	k	W_S	W_T
200	1	500	0
400	2	250	0
600	3	115	0
800	4	60	0
1,000	5	40	0
3,000	15	3	0.02
4,000	20	1.9	0.02
5,000	25	1.45	0.02
6,000	30	1.25	0.05
7,000	35	1.1	0.05
50,000	250	0	0.2
60,000	300	0	0.2
70,000	350	0	0.2

Table 4.1. *Weights for controller design ($\tau_d = 0.005$ which is the period of the repetitive signal)*

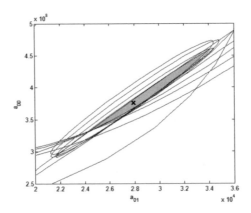

Figure 4.8. *Solution regions of the point condition*

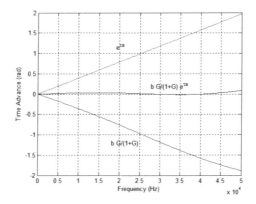

Figure 4.9. *Compensation of the phase lag by time advance*

4.2.6. Simulation results for the vertical direction

Numerical simulations are used to evaluate the repetitive controller designed for the vertical axis of the piezotube-based AFM. A square wave input of 40 nm height at 200 Hz is assumed to be the surface topography being scanned. The result obtained using a well-tuned PI controller for the vertical motion of the piezotube is illustrated in Figure 4.10a and the error is shown in Figure 4.10b. The oscillations occur due to the insufficient bandwidth of the system as was mentioned earlier.

Figure 4.11a illustrates the scan simulation under repetitive control instead of the PI controller and the corresponding error is shown in Figure 4.11b. Apparently, the scan obtained with the repetitive controller is better than the PI after the first two periods. The error is smaller at the moments of disturbances (step changes in surface topography) and so is the control effort as a consequence.

Another important factor in constant force scanning is the size of the interaction forces between the probe's tip and the sample surface. Large forces are not convenient to avoid probable damage on the tip and the sample, especially when the sample is made of organic matter like a biological specimen. The comparison of the interaction forces with the PI and the repetitive controller are demonstrated in Figure 4.12. After the first period, the forces are reduced considerably both on the flat parts of the steps and at the moments of disturbances when more control effort is needed.

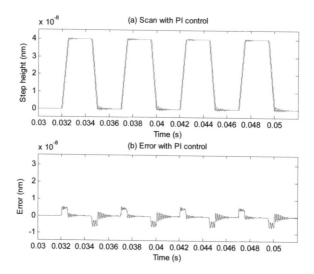

Figure 4.10. *Illustration of the scan with PI control (a) and the error on the probe's oscillation amplitude (b)*

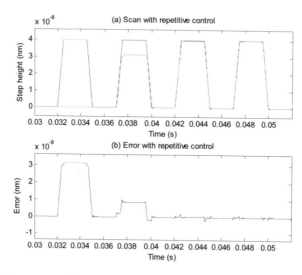

Figure 4.11. *Illustration of the scan with repetitive control (a) and the error on the probe's oscillation amplitude (b)*

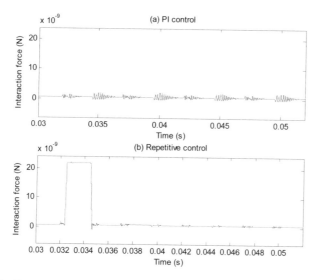

Figure 4.12. *Illustration of the interaction forces in the case of PI control (a) and the same forces in the case of repetitive control (b)*

4.3. MIMO disturbance observer control of the lateral directions

To illustrate different possible robust control methods that can be applied to AFM axes of motion, a disturbance observer is applied to the lateral scanning axes of a piezotube-based AFM in this section. Piezoelectric materials are widely used to obtain accurate motion of its axes, offering sub-nanometer precision. Tube-shaped piezoelements are commonly used as actuators because they are very good manipulators with their compact design, three degrees of freedom, low cost, and good properties for control purposes like stability, fast response, high bandwidth, and high precision. However, they also present undesired behavior like hysteresis, creep, thermal drift, and coupled motion of their axes [MOH 08], [DEV 07]. In particular, this coupled motion of the axes requires special attention when a piezoelectric tube actuator is used as a scanner for an AFM as it results in inaccuracies in absolute positioning and causes image distortions.

It is observed for most of the piezotubes that there is a coupling between the x- and y-axes, due to the inevitable eccentricity of the inner and outer cylinders of the tube as shown in Figure 4.13, which occurs during manufacturing. Moreover, there is a cross-coupling between the lateral and the vertical axes (not discussed here) which causes imaging inaccuracy when large areas are scanned. A piezostack actuated stage can be used as the nanopositioner instead of a piezotube to reduce the coupling effects [SAL 02]. However, it needs more complicated and expensive mechanical design to obtain three degrees-of-freedom motion. This complicated mechanical design may result in highly coupled motion of the axes while offering higher bandwidth [DON 07].

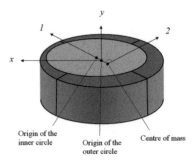

Figure 4.13. *x and y are the theoretical axes of motion whereas 1 and 2 are the actual axes of motion, the difference being caused by the eccentricity of the inner and the outer walls of the tube*

Previous work has been carried out in [TIE 04], [TIE 05] to compensate for the errors caused by the cross-coupling using inversion-based iterative control approach. However, the undesired motion in the lateral axes caused by the coupling between the x- and y-directions remains, as it causes image distortions in the $x–y$ plane and prevents absolute positioning for all sizes of the scan area.

[DAN 99] published their work that handles the lateral coupling problem by a MIMO controller with lateral feedback where a particular controller is designed for the ramp-type input signals used for triangular scan motion only. A more recent work is presented by Yong *et al.* [YON 10] on the lateral coupling problem of a piezostack actuated stage by using H_∞ controller design.

In fact, the coupling between the lateral axes has been customarily compensated for by introducing a variety of correction polynomials after an open-loop calibration process for commercial AFMs, rather than by using the above-mentioned control solutions. In this section, we propose a different method by offering an add-on digital control unit with an embedded algorithm that uses the *MIMO disturbance observer control* method of [AKS 10], for canceling the coupling between the lateral axes x and y of a piezotube by designing appropriate filters. One of the main contributions of the use of this control strategy is that the primary scans and iterations, which are not practical when manipulation of a particle instead of imaging of a surface is the case, are not necessary for learning purposes before imaging. Besides, this control strategy is not restricted to a certain type of input signal and hence can be used for motion profiles that have non-repetitive and/or unbiased trajectories. Moreover, the decoupling controller system proposed here is easy to tune.

An accurate model of the piezoelectric tube actuator is needed as a pre-requisite for control analysis and design. Previous work on linear models for ideal uncoupled lateral motion of the piezotube actuator has been published in [RAT 05] and an analytical model has been derived in [OHA 95]. A coupled analytical model has also been published in [RIF 01]. A fast, robust nanopositioning application has been treated in [LEE 09] using a two controller degrees-of-freedom LMI-based controller.

In this section, a MIMO model is used for the controller design. This MIMO model has been derived experimentally and contains the coupling effects between the lateral axes, as well. The organization of the rest of the section is as follows: in section 4.3.1, the experimental setup and the model of the quartered piezotube for lateral dynamics are given, whereas section 4.3.2 presents the basics and analysis of the MIMO disturbance observer algorithm used. In section 4.3.3, the design of a suitable controller and experimental results are given for the controlled and uncontrolled motions of the piezotube. The chapter ends with concluding remarks in section 4.4.

4.3.1. *The piezotube and the experimental setup*

The structure in Figure 4.14 is a custom built system where infrared sensors (Sharp GP2S60) are placed on the *x*- and *y*-axes for feedback. The infrared light is reflected from the body of the tube (Figure 4.1) onto a phototransistor output, which is a reflective photointerrupter with emitter and detector facing the same direction in a molding that provides non-contact sensing. It also blocks visible light to minimize false detection. The piezotube is placed on a granite plate to protect it against environmental vibrations. It is manipulated by a digital controller using the analog to digital converter (ADC), digital to analog converter (DAC), and a processor mounted on the digital control unit. A schematic representation of the system is sketched in Figure 4.15a and the corresponding picture of the setup is shown in Figure 4.15b. The calibration free decoupling algorithm (the MIMO disturbance observer) runs on the add-on digital control unit as shown in Figure 4.15a. The piezotube is placed in an acoustic enclosure to protect the whole AFM process against environmental sounds.

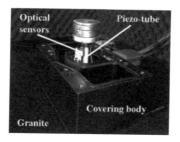

Figure 4.14. *Piezotube placed on granite with optical sensors measuring*
x- and y-axes motion

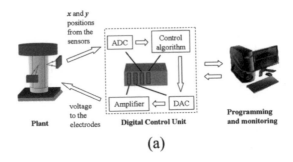

(a)

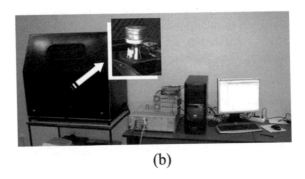

(b)

Figure 4.15. *(a) Schematic diagram and (b) the picture of the experimental setup and the hardware*

The dynamical behavior of the piezotube can be characterized as a well-known mass-spring-damper system, thanks to its structure. The sampling rate is 1 kHz because of the hardware conditions and discrete methods are used for system identification, analysis, and control. The whole system is modeled as given in equations [4.17]–[4.22]. The input is the digital value corresponding to the voltage output of the 18-bit DAC such that a unit increment equals 7.63×10^{-5} volts. Similarly, the output is the digital value corresponding to the voltage input of the 16-bit ADC such that a unit increment equal to 2.9×10^{-4} volts corresponds to a 10 nm displacement. The amplifier gain for the high voltage applied to the electrodes is 10.

The transfer functions in [4.19]–[4.22] are calculated by least squares curve fitting over the time responses of the system to various test signals, as they represent the direct throughputs and the undesired coupling effects, respectively. Note that the denominators of the direct throughputs [4.19] and [4.20] are essentially the same as expected because they arise from the same mechanical system. The denominators in the coupling effect transfer functions [4.21] and [4.22] are slightly different which is due to the fact that the coupled motion is small. Therefore, their estimation is not as accurate as in [4.19] and [4.20]. This situation does not constitute any problem since the coupling dynamics are only a measure of the model uncertainties to be robustly handled by the controller, and the direct throughput functions are essentially used for the controller design.

$$\begin{bmatrix} x \\ y \end{bmatrix} = \underbrace{\begin{bmatrix} G_{xx}(z) & G_{yx}(z) \\ G_{xy}(z) & G_{yy}(z) \end{bmatrix}}_{G(z)} \cdot \begin{bmatrix} u_x \\ u_y \end{bmatrix} \qquad [4.17]$$

$$G_{ij}(z) = \frac{j(z)}{u_i(z)} = \frac{b_{0ij}}{z^2 + a_{1ij}z + a_{0ij}} \quad \text{for } i = x, y \text{ and } j = x, y \qquad [4.18]$$

$$G_{xx}(z) = \frac{x(z)}{u_x(z)} = \frac{-0.0057}{z^2 - 1.5017z + 0.9331} \qquad [4.19]$$

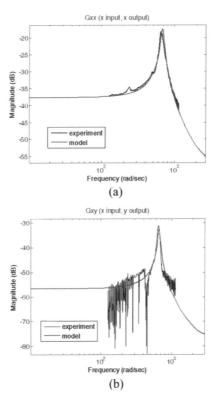

Figure 4.16. *(a) Frequency response from x to x, (b) frequency response from x to y. Similar plots are obtained from y to y and y to x as a function of applied field*

$$G_{yy}(z) = \frac{y(z)}{u_y(z)} = \frac{0.0044}{z^2 - 1.4995z + 0.9286} \qquad [4.20]$$

$$G_{xy}(z) = \frac{y(z)}{u_x(z)} = \frac{0.0006}{z^2 - 1.5480z + 0.9596} \qquad [4.21]$$

$$G_{yx}(z) = \frac{x(z)}{u_y(z)} = \frac{0.0005}{z^2 - 1.5160z + 0.9922} \qquad [4.22]$$

Superimposed plots of the frequency responses of the piezotube and the model functions, which are obtained from the frequency sweep up to 200 Hz and impulse responses, respectively, are given in Figure 4.16.

4.3.2. MIMO disturbance observer

The MIMO disturbance observer algorithm used for calibration free decoupled motion is presented in this section. The disturbance observer is a two degrees-of-freedom control system inside the control unit that makes an uncertain plant behave like its nominal or desired model. It also has excellent disturbance rejection properties. The coupling caused dynamics on a certain axis of the piezotube is treated as model uncertainty here and the MIMO disturbance observer is used to decouple the lateral axes of scanning. Consider plant P with multiplicative model error $W_m \Delta_m$ and external disturbance d. Note that the "y" term in the equations from [4.23] to [4.34] in the following discussion is the general output of the control system, not the y-axis. The input–output relation is expressed as follows:

$$y = Pu + d = (P_n(I + W_m \Delta_m))u + d \qquad [4.23]$$

where P_n is the nominal (or desired) model of the plant. P and P_n are square transfer function matrices of dimension two here and I is the identity matrix of dimension two. P_n is chosen as a non-singular and diagonal matrix as follows:

$$P_n = \begin{bmatrix} P_{n1} & 0 \\ 0 & P_{n2} \end{bmatrix} \qquad [4.24]$$

The aim of the disturbance observer is to obtain:

$$y = P_n u_n \qquad [4.25]$$

as the input–output relation which is called model regulation. Model regulation will allow the individual piezoelectric tube scanning axes to be decoupled as the desired model P_n in [4.25], with P_{n1} and P_{n2} in [4.24] being the decoupled x- and y-axes transfer functions, respectively. Doing this, we can achieve the desired response from the plant, that is, the x- and y-axes of the piezoelectric tube. The diagonal form of the transfer function matrix in the nominal model is critical for decoupling purposes since the original MIMO problem is divided into two independent SISO problems with this choice.

The extended error is defined by putting the external disturbance and the model uncertainty together and reformulating [4.23] as follows:

$$y = (P_n (I + W_m \Delta_m))u + d = P_n u + e, \ e = P_n W_m \Delta_m u + d \qquad [4.26]$$

which can be re-expressed as:

$$e = y - P_n u \qquad [4.27]$$

The effect of the extended error in [4.26] can be canceled using the following control law:

$$u = u_n - P_n^{-1} e = u_n - P_n^{-1} y + u \qquad [4.28]$$

which achieves the model regulation aim in [4.25].

However, P_{n1}^{-1} and P_{n2}^{-1} required when calculating the inverse of P_n in [4.28] are not causal and hence cannot be implemented. Therefore, we pre-multiply it by a low-pass filter Q in the following form:

$$Q = \begin{bmatrix} Q_1 & 0 \\ 0 & Q_2 \end{bmatrix} \qquad [4.29]$$

to make the diagonal elements of the 2×2 matrix QP_n^{-1}, namely Q_1/P_{n1} and Q_2/P_{n2} causal.

Including the sensor noise as well, $y + n$ is used instead of y for the actual output signal and the new control law becomes:

$$u = u_n - QP_n^{-1}(y + n) + Qu \qquad [4.30]$$

The block diagram of the above-mentioned control scheme is shown in Figure 4.17.

With this approach, two independent SISO loops can be designed with the nominal models P_{ni} and the Q_i filters with the above choice of [4.24] and [4.29] where $i=1, 2$. Usually, u_n is chosen to be a triangular wave on one axis while it is a ramp on the other axis to provide the raster scan motion.

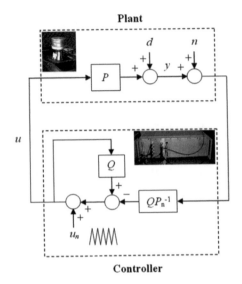

Figure 4.17. *MIMO disturbance observer architecture*

4.3.3. *Disturbance observer design for the piezotube and experimental results*

Note the following relation of the control scheme in Figure 4.17 between its inputs and output.

$$y = [I + P(I - Q)^{-1}QP_n^{-1}]^{-1}[P(I - Q)^{-1}u_n + d - P(I - Q)^{-1}QP_n^{-1}n] \qquad [4.31]$$

Using [4.31] and some manipulations, the input–output relation from u_n to y is as follows:

$$\frac{y}{u_n} = [(I - Q)P^{-1} + QP_n^{-1}]^{-1} \qquad [4.32]$$

The expression in [4.32] is equal to P when $Q=0$ and is equal to P_n when $Q=I$, which shows that model regulation is perfect for $Q=I$, the 2×2 identity matrix. Similarly, the input–output relation from d to y is:

$$\frac{y}{d} = I - P[(I - Q) + QP_n^{-1}P]^{-1}QP_n^{-1} \qquad [4.33]$$

and that is equal to the zero matrix when $Q=I$, which means perfect disturbance rejection. In addition to that, the input–output relation from n to y is:

$$\frac{y}{n} = -[I + P(I - Q)^{-1}QP_n^{-1}]^{-1}[P(I - Q)^{-1}QP_n^{-1}] \qquad [4.34]$$

Equation [4.34] is equal to the zero matrix when $Q=0$, which means perfect sensor noise rejection.

Since our aim is to decouple the motion along the lateral axes, the nominal models can be chosen such that $P_{n1} = G_{xx}$ and $P_{n2} = G_{yy}$ as given in [4.35].

$$P_{n1} = \frac{-0.0057}{z^2 - 1.502z + 0.9331}$$

[4.35]

$$P_{n2} = \frac{0.0044}{z^2 - 1.499z + 0.9286}$$

Considering the results above, Q_i, for $i=1$, 2 are chosen to be low-pass filters with unity DC gain which provide model regulation and disturbance rejection at low frequencies and sensor noise rejection at high frequencies. Basically, $Q_1 = Q_2$ and the poles are placed at 100 Hz and 120 Hz as in [4.36]. Thus, the frequencies within the operating frequency range of the piezotube, which is below its resonant frequency, are covered and the rest are filtered out. Figure 4.18 demonstrates the frequency response plots of P_{n1}, P_{n2}, and $Q_{1,2}$.

$$Q_i = \frac{0.247}{z^2 - 1.004z + 0.251}$$

[4.36]

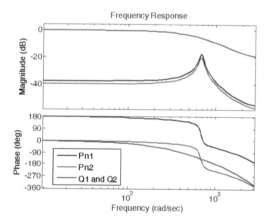

Figure 4.18. *Frequency response plots of the P_{n1}, P_{n2}, and $Q_{1,2}$ designed for the piezotube*

The corresponding difference equations are embedded into the digital control unit to implement the loop in Figure 4.17 using the designed filters. The following tests are carried out to see the results of the controlled and the uncontrolled cases.

Figure 4.19 demonstrates the decoupling ability of the MIMO disturbance observer controller for a step input in the x-direction. Although this input results in undesired coupled motion in the y-direction in the uncontrolled case, this undesired effect vanishes in the controlled case as illustrated in the experimental response of Figure 4.19. Figures 4.20 and 4.21 show the motion of the piezotube when rectangular and circular trajectories, respectively, in the x–y plane are desired. It is seen from Figure 4.20 that the rectangular trajectories are not coincident in the uncontrolled case. This is due to the hysteresis effect that is revealed significantly on the y-axis, which causes each rectangle to appear shifted on y (i.e. 2 in the coupled case, Figure 4.13). Besides, the x- and y-axes motions are rotated due to the eccentricity of the piezotube. Note that the simulation model represents the rotation of the axes (i.e. the coupling), but not the shifting of the rectangles because the hysteresis effect is not governed by this model. However, in the controlled case, the rectangles are not rotated thanks to the decoupled motion of the lateral axes and they are all coincident as desired. This shows that the undesired hysteresis effect is also handled by the MIMO disturbance observer successfully, as well as the coupling problem.

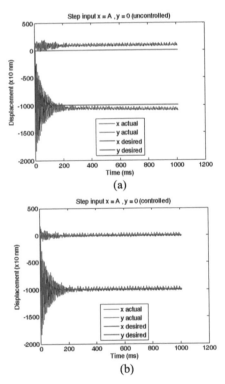

Figure 4.19. *Responses of the x- and y-axis in (a) the uncontrolled and (b) the controlled cases for a step input only on x*

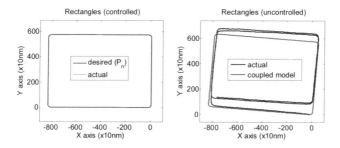

Figure 4.20. *Rectangular motion of the piezotube on x–y plane in (a) the uncontrolled and (b) the controlled cases*

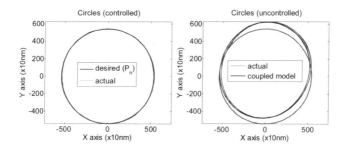

Figure 4.21. *Circular motion of the piezotube on x–y plane in (a) the uncontrolled and (b) the controlled cases*

The performance of the method is tested by a circular motion in the x–y plane, and the results are shown in Figure 4.19. In the uncontrolled case, it is seen that the center is not at the origin (zero) because of the hysteresis effect and the circles are shifted on y (i.e. 2 in the coupled case, Figure 4.13). However, the circles are just as desired in the controlled case. Here, the speed of the circular motion is 10 Hz.

An illustration of a sample scan is presented in Figure 4.22 to emphasize the difference between the coupled and the uncoupled imaging. The sample is artificially created on the computer as having square steps on its surface. The artificial sample surface consists of 550×550 points. A usual scan motion is experimented on the piezotube such that it is moved back and forth on x, while small steps advances are used on y.

The surface height (z position) corresponding to each x and y position of the piezotube are recorded for the illustration of the controlled and the uncontrolled cases. The results are plotted in 3D. It is seen from the figure that the image is distorted because of the coupling in the uncontrolled case. In addition, the heights of the square steps seem to be larger than they really are. Because the piezotube scans a smaller area than it should due to the hysteresis on *y*, the appearance of that area is stretched out during the imaging process (Figure 4.22).

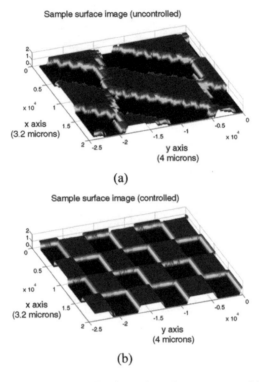

Figure 4.22. *Illustration of an artificial sample surface imaging in (a) the uncontrolled and (b) the controlled cases*

4.4. Concluding remarks

Robust control methods for the vertical and the lateral axes of a piezotube actuator used in an AFM were presented in this chapter. Control of the piezotube in the vertical axis and the lateral axes was treated separately to present two different approaches. Repetitive control was presented in combination with the vertical direction and MIMO disturbance observer control was presented in combination with the lateral axes. Simulations based on a realistic model were used to illustrate the results for the vertical axis and experiments were used for illustrating the results for the lateral axes. The COMES toolbox was used in both cases to fine tune the controller parameters (not shown here in the disturbance observer case for the sake of brevity) to satisfy mixed sensitivity-type performance specifications. The results presented demonstrate the applicability of these control techniques to AFM control. Note that the disturbance observer can be applied to the vertical axis also. Similarly, repetitive control can be applied to the lateral axes, preferably after MIMO disturbance observer compensation for better performance.

It is seen from the experimental results for the lateral axes that the MIMO disturbance observer works well for calibration free decoupling purposes. Hence, neither correcting factors nor pre-defined polynomials are required to avoid the distorted images. In addition, calibration of the absolute displacement of the axes is no longer necessary since they are forced to obey a nominal model and the desired response is pre-defined in that model. Therefore, the authors conclude that absolute positioning can be offered by having feedback on the orthogonal axes using appropriate sensors and controller hardware installed in the electronics of the conventional AFM.

Besides AFM imaging, the cancelation of the coupling of the axes is very important for manipulation of objects using a piezotube. In this case, the piezotube can be considered as a flexible robot arm where the motion along its axes is controlled satisfactorily by utilization of the MIMO disturbance observer algorithm in the digital control unit.

4.5. Acknowledgments

Section 4.1 that discusses the use of repetitive control in AFM including Figures 4.1–4.12 and Table 4.1 is taken from the ASME paper "Fast AFM Scanning with Parameter Space Based Robust Repetitive Control Designed Using The Comes Toolbox", by Serkan Necipoglu, Burak Demirel, Levent Güvenç, ESDA 2010 Engineering Systems Design and Analysis Conference, Paper Number ESDA2010-24499. The authors thank ASME for giving them permission to use that paper in this chapter.

The authors would like to thank TÜBİTAK, the Scientific and Technological Research Council of Turkey, for partial support for the second part of the work presented here through grant 108E256. The authors would like to thank Dr Ahmet Oral and Nanomagnetics Inc. for preparing the custom-made AFM piezotube used in the experiments here.

The authors would like to thank Prof. Cagatay Basdogan and the members of his research group at Koc University for cooperation and fruitful discussions on AFM and AFM control.

4.6. Bibliography

[ABR 07] ABRAMOVITCH D.Y., ANDERSSON S.B., PAO L.Y., SCHITTER G., "A tutorial on the mechanisms, dynamics, and control of atomic force microscopes", *Proceedings of American Control Conference*, New York, 2007.

[AKS 06] AKSUN GÜVENÇ B., GÜVENÇ L., "Robust repetitive controller design in parameter space", *ASME Journal of Dynamic Systems, Measurement and Control*, vol. 128, no. 2, 2006, pp. 406–413.

[AKS 10] AKSUN GÜVENÇ B., GÜVENÇ L., KARAMAN S., "Robust MIMO disturbance observer analysis and design with application to active car steering", *International Journal of Robust and Nonlinear Control*, Wiley, vol. 20, 2010, pp. 873–891.

[BIN 86] BINNIG G.K., QUATE C.F., GERBER C., "Atomic force microscope", *Physical Review Letters*, vol. 56, 1986, pp. 930–933.

[DAN 99] DANIELE A., SALAPAKA S., SALAPAKA M.V., DAHLEH M., "Piezoelectric scanners for atomic force microscopes: Design of lateral sensors, identification and control", *Proceedings of American Control Conference*, San Diego, CA, USA, 1999, pp. 253–257.

[DEM 09] DEMIREL B., Development of an interactive design tool for parameter space based robust repetitive control, M.S. Thesis, Istanbul Technical University, Istanbul, Turkey.

[DEM 10] DEMIREL B., GÜVENÇ L., "Parameter space design of repetitive controllers for satisfying a mixed sensitivity performance requirement", *IEEE Transactions on Automatic Control*, vol. 55, no. 8, 2010, pp. 1893–1899.

[DEV 07] DEVASIA S., ELEFTHERIOU E., MOHEIMANI S.O.R., "A survey of control issues in nanopositioning", *IEEE Transactions on Control Systems Technology*, vol. 15, no. 5, 2007, pp. 802–823.

[DON 07] DONG J., SALAPAKA S.M., FERREIRA P.M., "Robust MIMO control of a parallel kinematics nano-positioner for high resolution high bandwidth tracking and repetitive tasks", *Proceedings of 46th IEEE Conference on Decision and Control*, New Orelans, Louisiana, 2007, pp. 4495–4500.

[FUJ 08] FUJIMOTO H., OSHIMA T., "Nanoscale servo control of contact-mode AFM with surface topography learning observer", *10th IEEE International Workshop on Advanced Motion Control*, 2008, pp. 568–573.

[GAR 02] GARCIA R., PEREZ R., "Dynamic atomic force microscopy methods", *Surface Science Reports*, Elsevier Science B.V., vol. 47, 2002, pp. 197–301.

[GUN 07] GUNEV I., VAROL A., KARAMAN S., BASDOGAN C., "Adaptive Q control for tapping mode nanoscanning using a piezoactuated bimorph probe", *Review of Scientific Instruments*, vol. 78, 043778, American Institute of Physics, 2007.

[HAR 88] HARA S., YAMAMOTO Y., OMATA T., NAKANO M., "Repetitive control systems: a new type servo system for periodic exogenous signals", *IEEE Transactions on Automatic Control*, vol. 33, 1988, pp. 657–667.

[LEE 09] LEE C., SALAPAKA S.M., "Fast robust nanopositioning-a linear-matrix-inequalities-based optimal control approach", *IEEE/ASME Transactions on Mechatronics*, vol. 14, no. 4, 2009, pp. 414–422.

[LI 08] LI Y., BECHHOEFER J., "Feedforward control of a piezoelectric flexure stage for AFM", *Proceedings of American Control Conference*, Seattle, Washington, 2008.

[MOH 08] MOHEIMANI S.O.R., "Invited review article: accurate and fast nanopositioning with piezoelectric tube scanners: emerging trends and future challenges", *Review of Scientific Instruments*, vol. 79, 071101, 2008.

[OHA 95] OHARA T., YOUCEF-TOUMI K., "Dynamics and control of piezotube actuators for subnanometer precision applications", *Proceedings of the American Control Conference*, Seattle, Washington, 1995, pp. 3808–3812.

[ORU 09] ORUN B., NECIPOGLU S., BASDOGAN C., GÜVENÇ L., "State feedback control for adjusting the dynamic behavior of a piezo-actuated bimorph AFM probe", *Review of Scientific Instruments*, vol. 80, no. 1, American Institute of Physics, 2009, pp. 063701-1–063701-7.

[PAO 07] PAO L.Y., BUTTERWORTH J.A., ABRAMOVITCH D.Y., "Combined feedforward/feedback control of atomic force microscopes", *Proceedings of the American Control Conference*, New York, 2007.

[RAT 05] RATNAM M., BHIKKAJI B., FLEMING A.J., MOHEIMANI S.O.R., "PPF control of a piezoelectric tube scanner", *Proceedings of the 44th IEEE Conference on Decision and Control, and the European Control Conference*, Seville, Spain, 2005, pp. 1168–1173.

[RIF 01] EL RIFAI O.M., YOUCEF-TOUMI K., "Coupling in piezoelectric tube scanners used in scanning probe microscopes", *Proceedings of the American Control Conference*, Arlington, Virginia, 2001, pp. 3251–3255.

[SAL 02] SALAPAKA S., SEBASTIAN A., CLEVELAND J.P., SALAPAKA M.V., "High bandwidth nano-positioner: A robust control approach", *Review of Scientific Instruments*, vol. 73, no. 9, 2002, pp. 3232–3241.

[SAL 05] SALAPAKA S.M., DE T., SEBASTIAN A., "A robust control based solution to the sample-profile estimation problem in fast atomic force microscopy", *International Journal of Robust and Nonlinear Control*, vol. 15, 2005, pp. 821–837.

[SCH 01] SCHITTER G., MENOLD P., KNAPP H.F., ALLGÖWER F., STEMMER A., "High performance feedback for fast scanning atomic force microscopes", *Review of Scientific Instruments*, vol. 72, no. 8, American Institute of Physics, 2001.

[SCH 04] SCHITTER G., ALLGÖWER F., STEMMER A., "A new control strategy for high-speed atomic force microscopy", *Nanotechnology*, Institute of Physics Publishing, vol. 15, 2004, pp. 108–114.

[SCH 07] SCHITTER G., ASTRÖM K.J., DEMARTINI B.E., THURNER P.J., TURNER K.L., HANSMA P.K., "Design and modeling of a high-speed AFM scanner", *IEEE Transactions on Control Systems Technology*, vol. 15, no. 5, 2007.

[SRI 97] SRINIVASAN K., SHAW F.R., "Analysis and design of repetitive control systems using the regeneration spectrum", *ASME Journal of Dynamical Systems, Measurement and Control*, vol. 113, 1991, pp. 216–222.

[SUL 02] SULCHEK T., YARALIOGLU G., QUATE C.F., "Characterization and optimization of scan speed for tapping-mode atomic force microscopy", *Review of Scientific Instruments*, vol. 73, no. 8, American Institute of Physics, 2002.

[TIE 04] TIEN S., ZOU Q., DEVASIA S., "Control of dynamics-coupling effects in piezo-actuator for high-speed AFM operation", *Proceedings of 2004 American Control Conference*, Boston, MA, USA, 2004, pp. 3116–3121.

[TIE 05] TIEN S., ZOU Q., DEVASIA S., "Iterative control of dynamics-coupling-caused errors in piezoscanners during high-speed AFM operation", *IEEE Transanctions Control Systems Technology*, vol. 13, no. 6, 2005, pp. 921–931.

[VAR 08] VAROL A., GUNEV I., ORUN B., BASDOGAN C., "Numerical simulation of nano scanning in intermittent-contact mode AFM under Q control", *Nanotechnology*, vol. 19, 2008, pp. 075503-1–075503-10.

[WEI 97] WEISS G., "Repetitive control systems: old and new ideas", in BYRNES, C., DATTA, B., GILLIAM, D., MARTIN C. (eds), *Systems and Control in the 21st Century*, PSCT, Birkhäuser, Boston, 22, 1997, pp. 389–404.

[YON 10] YONG Y.K., LIU K., MOHEIMANI S.O.R., "Reducing cross-coupling in a compliant xy nanopositioner for fast and accurate raster scanning", *IEEE Transactions on Control Systems Technology*, vol. 18, no. 5, 2010, pp. 1172–1179.

Chapter 5

Automated Identification

This chapter addresses automated identification technology. In recent years, automated identification technology as part of mechatronics has become increasingly important in many areas such as industry, physical distribution, security, archiving, and medical application. Currently available and research-phase techniques related to identification technology are described in this chapter.

5.1. Introduction

We first look back at the historical development of the barcode and radio frequency identification technologies. In 1968, a barcode system for goods-wagon transport control was developed using a barcode symbol (Code 2 of 5) and a He–Ne laser scanner by Identicon Corp [HIR 01]. Afterwards, the serial barcode system grew into a system handling a large amount of information. By the 1980s (specifically 1984), the POS (point of sale) system using the JAN (Japanese article number) code was in practical use. Since then, the serial barcode system has been used to identify many articles such as foods, electrical appliances, medical apparatuses, medical materials, medicines, books, and clothes in distribution and physical distribution. In factories, barcode systems (using barcode Code 39 or NW-7) have been used in collecting production information and product instruction for realization of FA (factory automation) and CIM (computer integrated manufacture). In the late 1980s, several two-dimensional (2D) codes were developed in response to the requests by users for a barcode capable of handling much information in fields such as semiconductors, medicine, and

Chapter written by Hiroo Wakaumi.

mechanical components. Specifically, PDF417 and QR code were developed in 1989 and 1994, respectively. These barcodes including serial and 2D barcodes were standardized according to ISO (International Organization for Standardization) and IEC (International Electrotechnical Commission) in 1996 [HIR 01], and are now used in the field of sales management of articles in distribution, physical distribution, manufacturing, and individual identification.

On the other hand, the RFID (Radio Frequency Identification) system was developed to identify an IC tag attached to concrete-mixer vehicles to monitor the running status in 2002 [NIK 03, RFI 04]. This is a non-contact, non-directional identification system using a radio frequency wave and IC chips [RFI 04, JAP 03]. Presently, this system is used for personal identification, such as in the use of cash cards, credit cards, point cards, ticket cards, and commutation tickets, and in cattle identification.

Thus, identification technologies such as the barcodes and RFID have many applications. In recent years, automated identification technology as a part of mechatronics has become increasingly important because of the possibility of its wide use in many areas such as industry, physical distribution, security, archiving, and medical application. In this chapter, currently available and research-phase techniques related to this technology are described. In the following sections, conventional serial binary barcode and 2D barcode identification technologies are introduced. As a 2D barcode identification technology in a research phase, a multi-line scan based on the time-sharing laser light emission method using a laser is also presented. A newly developed ternary barcode detection technology currently in its research phase is then presented. Finally, RFID technology with many prospective applications is introduced.

5.2. Serial binary barcode

Serial barcodes such as Code 39, NW7, and Code 128 are binary barcodes consisting of white and black bars and spaces. Code 39 and NW7 are discrete barcodes consisting of the binary level with a wide bar or space (1) and a narrow bar or space (0). Code 128 is a continuous barcode consisting of multiple levels with four bar (space) widths. Techniques for scanning the serial binary barcode are a laser mechanical scan using a rotary mirror and an electronic scan using a CCD (charge-coupled device) line image sensor. These barcodes have the characteristics that they are easily read without contact and have a high identification rate when using the mentioned scanning techniques. The laser scanner and CCD scanner realize a nearly 100% identification rate.

Code 39 developed by Intermec Technologies Corp. (in 1975) was standardized in ISO and IEC in 1999. The code is currently used in physical distribution, product management in factories, and medical industry as main applications. One character of this barcode consists of five bars and four spaces, where three of the bars and three of the spaces are wide. This barcode can express 44 kinds of characters. The minimum symbol length is given by [5.1]:

$$L = (C + 2)(6X + 3N \cdot X) + G(C + 1) + 2Q. \tag{5.1}$$

Here, X, C, N, and G represent the narrow element width, the total number of characters not including start and stop codes, the ratio between wide and narrow elements, and the gap between neighboring characters, respectively. Q is the minimum width of the quiet-zone space, which is regulated at one larger than the width of the start and stop characters. When X is equal to or less than 0.508 mm, N is set to 2.2–3.0 (usually 2.5). The gap G is set to the largest of either $3 \cdot X$ or 1.35 mm. The height H is set to the largest of either 6.35 mm or 15% of all symbol length. Figure 5.1 shows a code pattern of one character of Code 39.

Code: 0 0 0 0   1   1   0 0   1

Figure 5.1. *Code pattern of one code 39 character*

NW7 developed by Monarc Marking in 1972 has characters consisting of seven elements, with each character composed of four bars and three spaces. Only two of the seven elements are wide. This barcode can express 20 characters. The barcode is used, for example, in clothing management, blood management, pick-up management for the home-delivery service, mail registration, lending management at libraries, and membership cards. The minimum symbol length L is given by:

$$L = \{(2N + 5)(C + 2) + (N - 1)(W + 2)\} X + G(C + 1) + 2Q. \tag{5.2}$$

Here, W is the total number of wide character codes. Figure 5.2 shows a code pattern of one character of NW7.

Code: 0 0 0 0   1   1   0

Figure 5.2. *Code pattern of one NW7 character*

Code 128 developed by Computer Identics Corp. in 1981 has characters consisting of three bars and three spaces, where each bar (space) has four types of bar (space) width. The code can express all 128 ASCII characters. Three types of code set – A, B, and C – are defined depending on the type of start code. The minimum symbol length L is given by [5.3]:

$$L = (5.5D + 11C + 35) X + 2Q \qquad\qquad [5.3]$$

Here, X and D represent the module width and the number of figures of the code set, respectively. C represents the total number of ASCII characters, function characters, shift characters, and code characters. Q represents the minimum quiet zone width, which is the largest of either $10X$ or 2.54 mm. This barcode being able to handle lots of information is used in the management of medical apparatuses, medical materials, and medicine. Figure 5.3 shows a code pattern configuration of Code 128.

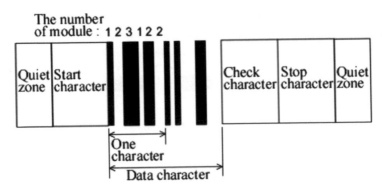

Figure 5.3. *Code pattern configuration of Code 128*

5.2.1. *Identification technology for serial binary barcodes*

We first describe the identification principle of the serial binary barcode. A barcode reader for identification of this kind of barcode consists of a light source, scanner, detector, amplifier, A–D convertor, decoder, and interface (Figure 5.4). A light-emitting diode (LED) or a laser diode is used for the light source. There are two self-scan-type techniques of CCD scanning and laser scanning for the scanner. The CCD has a self-scanning function in the device itself. However, in laser scanning, a polygonal mirror or galvano mirror is used for scanning laser light. The polygonal mirror has a highly stable rotation speed but is expensive, whereas the galvano mirror has an unstable scanning speed but is inexpensive. Therefore, for the latter, special signal processing techniques are required. A photodiode is used for the detector in laser scanning because of its high sensitivity. A photodiode array integrated in the CCD device is used in the CCD scanner. In the case of the A–D convertor, sophisticated techniques are required to obtain the digitized bar and space signals corresponding to each bar and space width. The decoder measures barcode patterns as width information by counting clock pulses. These measured patterns are classified as 1 or 0 corresponding to wide and narrow bar or spaces. The pattern data are decoded as a character or sign and converted to ASCII character codes. Subsequently, these are recorded in the memory. Finally, in the interface, memorized data are sent to an RS-232C interface or a TTL serial interface on PCs or terminals.

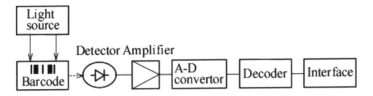

Figure 5.4. *Identification principle of the barcode*

5.2.1.1. *CCD scanning*

A light source of red LEDs is used to illuminate uniformly the barcode. The wavelength of red LEDs used is mainly 660 nm. The reflected light from the barcode is detected by a CCD line image sensor that consists of 2,000–3,000 elements [SEQ 75]. The output signal of the CCD sensor is converted by the A–D conversion circuit and its digitized signal is decoded in the decoder. The configuration of the CCD line image sensor is shown in Figure 5.5. This sensor is able to detect optical signals away from the object both by contact and remotely. In the former case, the reading width of the barcode is usually narrower than that of the scanner head. That is typically 80 and 100 mm, which are available for use in applications of FA and physical distribution. This scanner can be made cheaply and has a high identification rate. In the latter case, remote identification using a CCD sensor with 10 cm focal distances is possible at several centimeters to 10 cm from the object. Because wide barcodes can be detected using this sensor, there is no need to prepare scanners corresponding to the barcode width. That is, the application range is expanded.

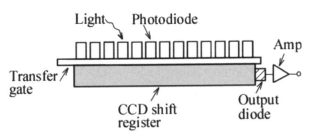

Figure 5.5. *A configuration of the CCD line image sensor*

5.2.1.2. *Laser scanning*

Basically, a laser scanner (using a red laser diode) is used in handheld scanners. The wavelength of laser diodes, which are most frequently used for barcode readers, is 660–680 nm. Laser light reflected by a rotary mirror with multi-reflection mirror facets scans a barcode surface as shown in Figure 5.6 [WAK 10a]. The light reflected from the barcode is detected by a sensor through the optical filter and optical–electrical converted signals are identified by a decoder. The scanning speed is 30–50 scan/s. The detection resolution changes depending on the detection distance. Usually, the resolution decreases as the scanner moves further from the focal point and nearer the one. Remote identification, large detection width (long detection depth), and maintenance of the high identification rate are features of this scanner. Therefore, this scanner is useful for a POS system and goods management in physical distribution.

Fixed-type laser scanners are those that are able to read barcodes automatically. The principle of reading barcodes is the same as that of handheld scanners. In the case of a fixed device, scanners and barcodes are positioned such that the reading and start of detection are automatic. The feature of this scanner is its high-speed detection. The scanner can operate at scan speeds higher than 400–2,000 scan/s. In addition, this scanner has features of long detection distance, long detection depth, and large detection width. Therefore, it is used on FA production lines and in physical distribution.

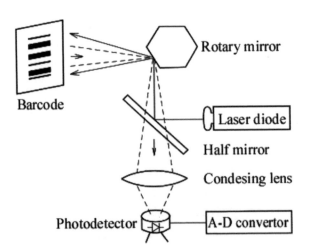

Figure 5.6. *Schematic outline of a laser scan system*

5.2.2. Requirements for serial binary barcode identification

The identification rate is affected by the contrast of barcode symbols. The print contrast signal (PCS) and space reflection rate R_s as indices of contrast are defined by the ANSI (American National Standards Institute) as follows:

$$\text{PCS} = (R_s - R_b)/R_s \times 100 \% \geq .75 \% \tag{5.4}$$

$$R_{smin} \geq 50 \% \text{ (for } X < 0.51 \text{ mm)} \tag{5.5}$$

$$R_s/R_b \geq 4 \tag{5.6}$$

Here, R_s, R_b, and R_{smin} represent the reflection rate of space, reflection rate of a bar, and minimum reflection rate of space, respectively.

5.2.3. Decoding for identification

The identification rate depends on factors such as the quality of a barcode label, optical resolution of a barcode reader, A–D convertor, and decoding algorithm. In the case of the decoder, the time period of the bar or space pattern signals measured as the number of clock pulses is classified as 1 or 0 compared with a reference value. The reference value cannot be constant because the time period changes depending on the scanning speed and the bar width. Usually, after the initial bar time period is measured, the reference value is defined. Next, the reference value is modified depending on the following bar width measured. Using this method, the system can be subordinated to the scanning speed. Thus, the identification rate increases.

Generally, the identification rate is higher for high-speed scanners. Scanners such as CCD scanners and laser scanners scan 10 times for identification. The scanner's reading rate per single scan is low, but the overall reading rate increases because of the multiple scanning. However, the first reading rate is important. When the first reading rate FR is too low, complete identification is impossible (Figure 5.7). This is seen from the dependence of the identification rate P on the FR that changes as shown in formula [5.7].

$$P = 1 - (1 - FR)^N \tag{5.7}$$

Here, N is the number of scans.

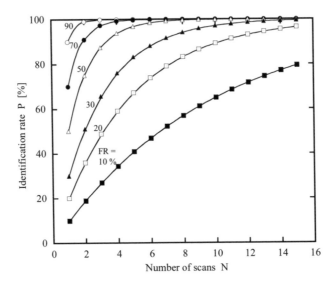

Figure 5.7. *Identification rate versus number of scans. FR represents the first reading rate*

5.3. Two-dimensional binary barcode

PDF417 developed in 1989 and QR code developed in 1994 are typical 2D barcodes. PDF 417 is a stack-type 2D barcode that can handle binary codes in addition to ASCII characters and numerals (Figure 5.8). This barcode consists of a basic module (a code word) of 17 modules with four bars and four spaces.

Figure 5.8. *A symbol example of PDF417 [HIR 03]*

The symbol consists of a start code, a left indicator, data, a right indicator, and a stop code (Figure 5.9). The column number of the code word is 1–90 and the maximum number of lines is 3–90. The left and right indicators include a line number, a total line number, a total column number, and a security level. The first code word in the data represents the total number of code words and the last code word an error detection code word. The symbol length L including a quiet zone is given by:

$$L = (4 + 17R)\,X. \hspace{4cm} [5.8]$$

Here, R is the number of code words per line. The quiet zone is set to more than twice the module width X. The minimum module height is set to more than three times the module width X. When X is 0.191 mm, the minimum module height is 0.57 mm.

This barcode with a large number of columns is identified using raster scanners or CCD scanners. Since this barcode encoded with much information is easily identified by usual barcode readers, it is used most widely around the world.

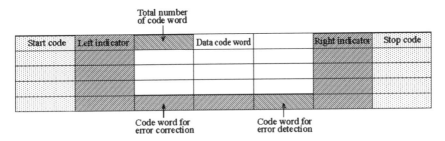

Figure 5.9. *Configuration of PDF417*

On the other hand, QR code is a matrix-type code that has large cells and view finders on three corners (Figure 5.10). Owing to this structure, the reference position, origin, symbol size, and tilt are detected at high speed. When the number of cells, cell size, and one-sided quiet zone width are N, X, and Q, respectively, the vertical or horizontal length of the symbol L is given by:

$$L = N \cdot X + 2Q. \hspace{4cm} [5.9]$$

This code that decodes the cell's layout pattern is difficult to read by scanning. It is instead read by area image readers. A QR code with less than 100 characters consisting of English letters and numerals can be read within 32 ms.

Figure 5.10. *A symbol example of QR code [HIR 03]*

5.3.1. *Scanning technology of the 2D barcode*

An electronic scan using a CCD or MOS area image sensor (area solid image sensor) is a barcode scanning technique for the 2D binary code such as PDF417 and QR code. In image sensing techniques, light emitted by the light source of LEDs illuminates 2D codes uniformly. In this case, uniform illumination using a reflection plate is important for the subsequent processing of the image signal. The 2D binary code image per field is stored in an area solid image sensor using a focusing lens. This image transformed to electronic signals is read out and stored in the memory (Figure 5.11). Through image processing, the type of 2D code, position, size, tilt, origin, and distortion are detected. The image data are encoded as optimal data for obtaining efficiently encoded symbol data. The white (1) or black (0) information for each cell from the optimal data is determined through comparison with a threshold. The digital data obtained by comparison is transformed to a corresponding symbol character using a symbol character table. When using area image sensors, a decoding time or speed is determined by how image processing is done. Optical focusing is also important for obtaining a sharp image. An automatic remote focusing method using both an LED and a laser diode is used recently. Area solid image sensors can read 2D codes and stack-type barcodes in addition to binary barcodes. Only stack-type barcodes can also be read using laser raster scanners.

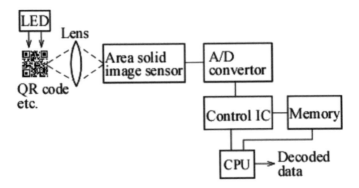

Figure 5.11. *Schematic outline of an area solid image sensor scan system*

5.3.1.1. *CCD area solid image sensor*

There are mainly two types of CCD area solid image sensors: which are a frame-transfer device with an imaging area and an additional transfer area, and an interline-transfer device integrating these areas within the same area [SEQ 75]. The configuration of a CCD area solid image sensor of the interline-transfer type suitable for a compact integration structure is shown in Figure 5.12. While this configuration enables simple operation of imaging and storage functions, it results in a complex cell design and a reduction of the light-sensitive area. The operation principle of the sensor is described below. When an optical image is illuminated on the photodiode, the charge is accumulated in the separate photodiodes. After the charge in each cell is stored in the previous field, it is transferred to the vertical shift register and then shifted to the horizontal shift register by one element during each horizontal period. All shifted charges are then transferred to the output diode along the horizontal shift register. The charge is transformed to a voltage signal at the output diode and the amplifier. Thus, a conversion of optical image of barcode patterns to electrical signals is achieved. The CCD solid image sensor has features of less noise and higher sensitivity compared with the MOS area solid image sensors described in the following section.

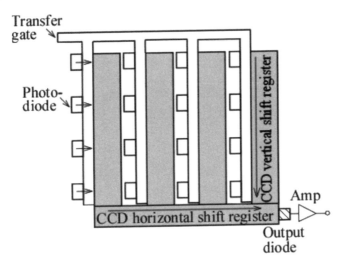

Figure 5.12. *Configuration of a CCD area solid image sensor*

5.3.1.2. *MOS area solid image sensor*

A MOS area solid image sensor is also used as a scanning device. The configuration of this structure is shown in Figure 5.13. This device has a structure with horizontal/vertical shift registers and integrating in a x-y matrix style each pixel which consists of a photodiode and vertical MOSFET switch [TAD 03]. The incident light is illuminated to the photodiode. Pulses from the horizontal/vertical shift registers are sent sequentially to the vertical/horizontal selection lines. When gate pulses from the horizontal and vertical registers are given to the gates of the vertical MOS FET switches and a column selection switch, the charge on the photodiode in the pixel selected is read out by discharging through the selected vertical MOSFET switch and the column selection switch to the output amplifier.

Currently, the advanced version of this sensor reduces the reset- and dark-current noise and increases sensitivity using a buried photodiode, and is used as a CMOS area solid image sensor for digital video and single-lens reflex cameras. Such devices feature easy integration of peripheral circuits and lower power dissipation but a lower signal-to-noise ratio compared with the CCD area solid image sensor.

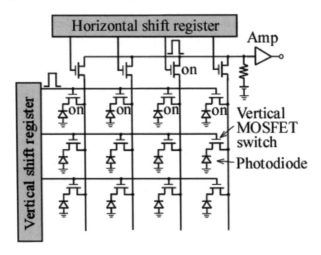

Figure 5.13. *Configuration of a MOS area solid image sensor*

5.3.2. *Multi-line scan based on time-sharing laser light emission*

Goods identification in real time is attractive in applications such as goods management in production lines and automated warehouses requiring high-speed detection. To realize real-time identification, the barcode needs to contain much information. Although the use of high-density 2D binary barcodes or color barcodes for identification systems has been considered, the scanning speed in these techniques is limited to nearly 50 scan/s by the complicated image processing using CCD/MOS area solid image sensors, and this makes it difficult to achieve the high-speed sorting of goods. Furthermore, because the depth of the field in a camera lens using CCD/MOS area solid image sensors is shallow (< 2 cm), the focus needs to be adjusted to realize a longer detection range. For high-speed identification of stack-type 2D barcodes such as PDF 417, raster laser scanners are appropriate. In conventional simple laser scanners, a physical scanning speed is determined by the rotation speed of the polygonal-mirror scanners. This simple type of scanner is insufficient for applications in which high-speed identification is necessary, because the scanning speed is limited by the scanner's physical characteristics. Hence, the key to achieving a high-speed identification system is to ensure that the system can identify a large amount of barcode information within a single scanning period.

5.3.2.1. *Three-line scan by time-sharing laser light emission*

For achieving high-speed identification, a multi-line-scan 2D binary barcode detection system (BCDS) using multi-LDs with time-sharing light emission (TSLE) operation for a stack-type 2D barcode was proposed in 1999 in the research phase [WAK 99, WAK 00b, WAK 01]. This enables multi-line barcodes to be read within a single scanning period. The 2D binary BCDS employing the TSLE operation is able to detect a stack barcode consisting of n lines within a scanning period of one facet of a polygonal mirror. Here, a three-line BCDS capable of realizing this is described for demonstration. Figure 5.14 is a schematic outline of the detection system, which consists of three LDs to detect a three-line barcode. Figure 5.15 shows waveforms to explain the operation principle. In this system, light emitted from the three LDs is reflected on to a polygonal mirror and reflected light from the mirror scans each line of the barcode in a time-sharing mode at a high LD drive frequency. Scattered reflected light from the three-line barcode is detected by a single photodetection apparatus consisting of a photodiode, photodetection lens, and a photodetection amplifier. The detected output is sampled in the time-sharing mode using sampling pulses synchronized with LD drive pulses. Three sampled outputs are digitized in the A/D conversion circuit. One of these digitized outputs (CH2*) is delayed for a barcode signal period T_0 and another (CH3*) is delayed for twice the barcode signal period $2T_0$ in the delaying circuit. Two delayed outputs (CH2** and CH3**) and another non-delayed digitized output (CH1*) are added together in the summing gate. The added outputs are decoded successively in a decoder within a scanning period (T_0), which contains a blanking period. Therefore, with this

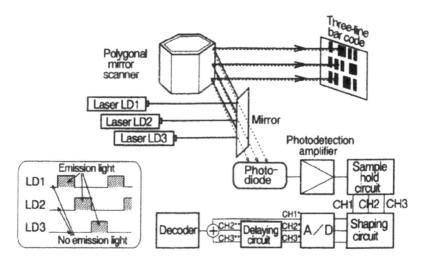

Figure 5.14. *A schematic outline of the 2D binary BCDS with the TSLE consisting of three LDs and a three-line barcode [WAK 01]*

operation, the scanning of three barcode lines is completed within a period of one facet of a polygonal mirror; effective high-speed scanning triple that in conventional 2D barcode scanners is possible. Use of a single photodetection apparatus enables a scanned barcode surface to achieve wide defocus because there is no need to focus on photodetection apparatus arranged in different spatial positions.

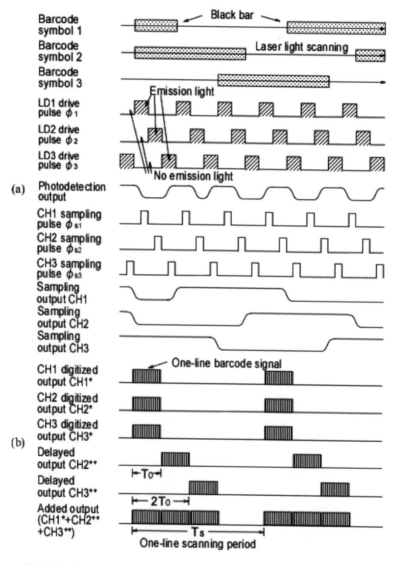

Figure 5.15. *Waveforms in the 2D binary BCDS employing the TSLE operation to explain operation principle. T_0 represents a one-line barcode signal period [WAK 01]*

5.3.2.2. *Optimization of LD bias current*

When visible light LDs are used, we must take care in a drive mode of the LDs. As shown in Figure 5.16, the conventional modulation drive of index-guided-type LDs without bias current achieves a rise time *tr* of no less than 0.9 µs owing to its slowly rising light emission phenomenon [WAK 96, WAK 98a, WAK 99]. This phenomenon of slowly rising light emission is not caused by the laser driver circuit or packaging. *tr* just below 100 ns is required to achieve an LD drive frequency of 2 MHz. In the TSLE system, it is desirable that LDs are driven using the bias current as predicted from the current-light output characteristics of Figure 5.16. However, a bias current corresponding to a bias light must be optimized to a suitable one because there is a tradeoff between the high-speed pulse modulation operation and the threshold of LD (*I*th ~31 mA). Channel crosstalk between three-line barcode signals caused by the bias light was tested. Channel crosstalk test results showed that a bias current below 30 mA is desirable to obtain small channel crosstalk between two channels. On the other hand, a bias current near the threshold of LD (~28.5 mA) is desirable to achieve a high-speed light output rise time of 90–130 ns (Figure 5.16). Therefore, the optimum bias current *I*opt is 28.5 mA, which is slightly less than the threshold of LD (*I*th ~31 mA). Thus, a pulse modulation drive with an LD bias current near the threshold (PMBC drive) developed in 1996 is useful for achieving high-speed switching of LDs [WAK 96, WAK 98a, WAK 99]. This method enables high-speed light emission, unlike the conventional modulation drive method of LDs. Hence, it enables high-frequency driving of LDs; that is, high-speed scanning.

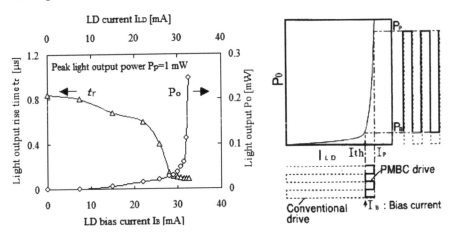

Figure 5.16. *(a) Light output rise time versus LD bias current, light output versus LD current characteristics. (b) Light output pulse in relation to the magnitude of LD bias current*

5.3.2.3. *Experimental results*

A prototype system consisting of three LDs for a three-line barcode has demonstrated the effectiveness of the TSLE operation (Table 5.1). The interval between neighboring LD light paths is 9 mm. The interval l_1, l_2, and l_3 between laser diodes LD1, LD2, and LD3 and the mirror are 197, 103, and 16 mm, respectively. This interval has been designed to be different from each others because of using spatially overlapping LD heads to make the LD light paths' interval as small as possible. The distance between the polygonal mirror and the mirror was set to 60 mm. The polygonal mirror used is a six-mirror-facet polygonal scanner with the size of one mirror facet being 25×25 mm^2. Index-guided type, 680 nm, LDs in the system were used because of small astigmatism. A photodetection apparatus consists of a photodetection lens with an aperture of 20 mm φ, a Si PIN photodiode (S3071), and a three-stage amplifier with a bandwidth of 6.4 MHz and a gain of 71.54 dB. Barcodes consist of three-line Code 39 symbols with a minimum width of 0.4 mm and a length of 44 mm arranged in parallel.

Photodetection amplifier	Bandwidth of 6.4 MHz, Gain of 71.5 dB
Photodiode	Si PIN photodiode, Capacitance of ~15 pF
Rotating mirror scanner	
·Number of mirror facets	6
·Size of one mirror facet	25×25 mm^2
Laser diode	Index-guided type, 680 nm
Lens	20 mmφ
Space between lights emitted from two LDs	9 mm
Barcodes	Code 39 symbols with length of 44 mm arranged in parallel, a minimum bar width $W = 0.4$ mm

Table 5.1. *Specifications of the prototype 3-line BCDS with TSLE operation*

In this system, the pulse modulation method with a bias current near the threshold (PMBC drive) of index-guided-type visible light LDs was used to improve laser light output rise time (t_r) characteristics. For example, the optimum bias current of 28.5 mA for LD2 was chosen to achieve a high-speed light output rise time t_r of ~130 ns. The system could recognize the barcode symbol pattern with a minimum width of 0.4 mm under the condition of one-line scanning period 2.05 ms for $L =$ 10 cm. Figure 5.17 shows the maximum effective scanning speed versus LD drive frequency. The maximum effective scanning speed V_{es} was estimated under the condition that barcodes could be recognized within the practically usable detection lengths (approximately 7–17 cm). The achieved V_{es} was 1,460 scan/s when the drive frequency f_c was 2 MHz. This is two and nine tenths times the speed of conventional

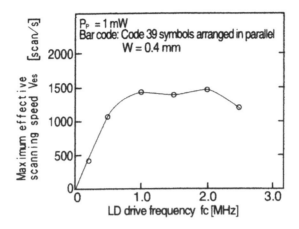

Figure 5.17. *Maximum effective scanning speed versus LD drive frequency in the 2D BCDS with the TSLE consisting of three LDs and the three-line barcode [WAK 01]*

2D barcode scanners. A decrease in V_{es} for f_c between 1 and 2 MHz is due to the wide interval between the neighboring sampled points in one barcode symbol. A sampled barcode signal waveform with a wide interval between neighboring sampled points becomes disarranged by noise within a non-sampled period and rough, when f_c is somewhat high but there are few sampled points. As a result, barcode recognition in f_c of ~1.5 MHz cannot be achieved at a high scanning speed of ~1,440 scan/s. That is, the scanning speed enabling us to recognize barcodes must be less than ~1,440 scan/s to increase the number of sampled points. An increase in V_{es} for f_c = 2 MHz is due to an increase in the number of sampled points for one barcode symbol. V_{es} is limited by the narrow bandwidth characteristics of the photodetection amplifier. In addition, a long detection range (~10 cm) for the system was achieved without the need to adjust the focus, unlike the case for cameras when using CCD/MOS area image sensors.

Furthermore, a 2D BCDS with mask collimators was proposed to control the aperture of the laser beam to realize high detection resolution [WAK 04].

A system consisting of masks with 2.5 and 3 mm apertures was confirmed to be capable of providing sufficient resolution for 0.25 mm barcodes over a detection range of more than 12.5 cm.

5.4. Ternary barcode

In a 2D binary BCDS using a raster laser scanner appropriate for relatively high-speed detection, the amount of information handled by barcodes is limited by the number of raster scanner mirror facets and the detection resolution of the identification system. In addition, in a 2D BCDS with TSLE, the number of barcode lines is limited to nearly four because of the photodetection amplifier bandwidth limitation of about 6.4 MHz realizable in practical use (that is, the number of sampling points is limited), and thus, not so much information can be handled [WAK 01].

Therefore, a ternary barcode with much information was devised in 2005 [WAK 05, WAK 06a]. Although laser scanners are now being used to identify binary barcodes in production lines and elsewhere, no laser scanner enabling the system to identify half-tone barcodes has been developed yet. However, because conventional laser scanners or raster scanners can be used in detecting this kind of barcode through the development of new signal processing methods, there is the possibility of operating the detection system at high speed.

5.4.1. *Dual-threshold method*

In this section, a ternary barcode detection system (TBDS) with the laser scanner employing a dual-threshold (DT) detection method is described, providing a long-range detection range without having to adjust the focus. It also provides a high scanning speed while maintaining a great amount of information.

5.4.1.1. *Dual-threshold ternary barcode detection system*

An outline of the DT TBDS using a laser diode scanner is shown in Figure 5.18, and the detection method is explained in Figure 5.19. This system obtains an enveloped line of a detected barcode signal using an enveloped line detector and obtains two reference signals by decreasing this enveloped line to two levels with 60% and 25% attenuators. These reference signals, subtracted from the original barcode signal in different subtraction gates, obtain low-average and high-average signals. Thus, a bent barcode signal with a hyperbola-shaped enveloped line is easily changed to unbent low- and high-average signals. Gray–black mixed and black code signals are obtained by comparing these low- and high-average signals with independent thresholds V_{T1} and V_{T2} in gray-level and black-level comparators, respectively. This method automatically allows the system to control comparison levels to a proper level depending on the barcode signals. Therefore, code signals can ideally be obtained regardless of the detection distance. The mixed code signal is delayed by a time difference Δt between the gray–black mixed and the black code signals at each rise or fall change (caused by the difference in comparison levels) to obtain a selection pulse and produce a gray code signal. The selection pulse is created using a sampling circuit, which samples the black code signal selectively at the falling edge of the delayed mixed code signal. When the black code signal is low, the selection pulse becomes low at the first pulse of the black code signal, and this state is maintained until the high state of the following black code signal is sampled. Therefore, the gray code signal is obtained by deleting only signal parts corresponding to black bars using the logical product gating operation of the selected pulse and the delayed mixed code signal. The gray and black code signals are then added together in the summing gate, and the added output is decoded in a decoder, which chooses the gray or black code signal with the selection pulse. This system does not need complicated processing. Therefore, it is expected to detect barcodes at higher speeds and over longer ranges.

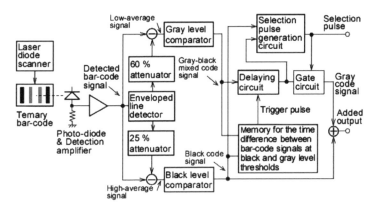

Figure 5.18. *Schematic outline of the dual-threshold ternary barcode detection system [WAK 06a]*

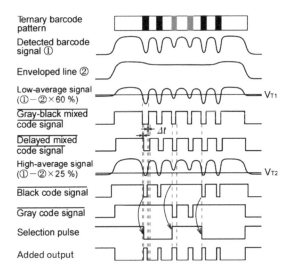

Figure 5.19. *Operation waveforms in the ternary barcode detection system employing the dual-threshold detection method [WAK 06a]*

When the ternary barcode is used, the amount of information greatly increases compared with that of a conventional binary barcode. In the binary barcode of Code 39, the maximum number of expression characters is 44. However, a combination of ternary barcodes consisting of nine elements per character enables the system to express at least 640 characters. This means that over 14 times more information can be realized using the ternary barcode, which enables the expression of numerals and Chinese characters, in addition to the basically required alphabet.

The attenuation degrees of two attenuators are set at 0.6 and 0.25, considering the condition in which the longest detection range could be achieved. The optimum thresholds V_{T1} and V_{T2} are set at 0.1 and 0.05 V, respectively, considering the maximum detection range.

5.4.1.2. *Experimental results*

The prototype system was tested. The detectable range for a minimum bar width W of 0.4 mm for a barcode was over 7.5 cm (Figure 5.20), which is practically usable. W is limited by a decrease in the changes of the barcode signals from a narrow white to a narrow gray or narrow black bar. In addition, as a test result of the possibility of high-speed detection, a maximum scanning speed of 370 scan/s, which is seven times the ~50 scan/s achieved by conventional CCD cameras, was obtained under the practical detection range ($\geq$4 cm).

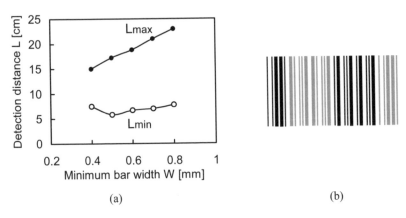

(a)   (b)

Figure 5.20. *(a) Detection distance versus minimum bar width [WAK 06a]. The scanning speed is 333 scan/s. (b) The ternary barcode used in the experiment*

Employing DT techniques, two-line and six-line ternary barcode identifications were also investigated to clarify the possibility of high-speed multi-line detection. A two-line DT BCDS was tested in combination with the TSLE technique and shown to be capable of detection at a scanning speed over 16 times that of conventional CCD cameras [WAK 06b]. A six-line DT BCDS was tested by combining with a raster scanner and shown to be capable of detection at a scanning speed of 417 scan/s, half that of the two-line system [WAK 07, WAK 09]. The six-mirror-facet raster scanner with a unit angle $\theta = 0.61°$ (slant angles of six mirror facets = $+2\theta$ to -3θ) and dimensions of one mirror facet of 15×25 mm^2 was used in this experiment. The experiment result for the slant angle of the barcode showed that the allowable skew angle for detection of the barcode is $-16°$ to $+16°$, which is nearly equivalent to that for the conventional 2D BCDS.

5.4.1.3. Theoretical study of the multi-line barcode configuration

In a 2D BCDS using a tilted mirror scanner (raster scanner), the position of a laser scan beam changes depending on the detection length. Therefore, laser scanning beam traces need to be made clear in designing an optimum BCDS. The vertical deviation of laser scanning beams depends on the incident angle at which the beams are incident to the tilted mirror facet, $\varphi/2$ [WAK 98b, WAK 00a]. According to Li's scan pattern theory [LIY 95], a vertical deviation x of the laser scanning beam reflected from the tilted mirror referring to the position of the beam reflected from the non-tilted mirror facet is given by:

$$x = L \cdot \tan(2n\,\theta) \cdot \cos(\phi/2)\ (n = 1, 2, \dots). \tag{5.10}$$

Thus, the scanning beam shifts far from the orthogonal direction as the incident angle increases. This means that as the laser beam deviates from the central part of the barcode, the vertical deviation x either increases or decreases. This model allows an optimum 2D BCDS including an optimum barcode configuration and its position to be designed easily.

Using this model, a usable barcode configuration was studied theoretically. In the prototype raster scanner, the tilt angles of the six mirror facets are set at values between -3θ and $+2\theta$. That is, three mirror facets are tilted in the lower direction and two mirror facets in the upper direction. The relation at this time between the location of a laser scanning beam and position of the barcode is shown in Figure 5.21. The lowermost laser scanning beam reflected from the scanner tilted to the lowest direction is located at $L \cdot \tan(6\theta) \cdot \cos(\varphi/2)$. The boundary position of the lowermost barcode and the outside is $3.5h$ (where h is the barcode height per barcode line). The boundary position of the second lowermost barcode and the lowermost one is $2.5h$. Considering the severest condition when the laser scanning beam departs from the barcode pattern, the condition in which laser scanning beams can pass through each line of barcode is as follows:

$$2.5h < L \cdot \tan(6\theta) \cdot \cos(\varphi/2) < 3.5h \tag{5.11}$$

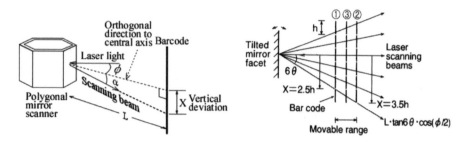

Figure 5.21. *(a) Vertical deviation of the laser scanning beam depending on the incident angle of the laser beam. (b) Relationship between laser scanning beams and the six-line barcode position [WAK 00a]*

This relation is presented as the detection length versus one-line barcode height in Figure 5.22. Practically usable barcode heights are the dotted and lined areas encircled by the detection length limits shown in the figure. $Lmax^*$ is mainly determined by the signal to noise ratio of the signal processing circuits and $Lmin^*$ is limited by the size of the scanner apparatus installed with the tilted mirror scanner. In the figure, as the incident angle $\varphi/2$ decreases, the usable barcode height extends. For example, when the range above 3 cm of L is allowed, the optimum barcode height extends from 1.4–2.7 mm at $\varphi = 90°$ to 1.6–3.1 mm at $\varphi = 71°$.

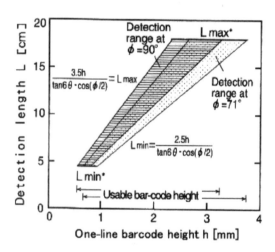

Figure 5.22. *Detection length versus one-line barcode height [WAK 00a]*

5.4.2. *Envelope differential composite method*

In the DT TBDS, the detection resolution is limited to 0.4 mm because of distortion of the detection signal from an uneven signal change that depends on the bar width. This distortion is due to the detection signals corresponding to narrow bars (black, white, and gray bars) in high-density barcodes decreasing in amplitude to low levels because of the decrease in the amount of laser light reflected from the barcodes. In this section, the TBDS employing an envelope differential composite method (proposed in 2008 [WAK 08]) is shown to provide high barcode densification with long range and high-speed scanning of a great amount of information.

5.4.2.1. *TBDS employing the envelope differential composite method*

An outline of the envelope differential TBDS using a laser diode scanner is shown in Figure 5.23. The detection method is explained in Figure 5.24. A slightly enveloped line of a detected barcode signal is obtained using an enveloped line detector with a small discharge time constant τ and a reference signal by decreasing the enveloped line intensity to a level with 60% attenuators. The use of the slightly enveloped line is indispensible for suppression of the detection signal distortion in the detection of high-density barcodes. An average signal is obtained by subtracting the reference signal from the original barcode signal with a subtraction gate and amplifying with an amplifier with a gain of 1.3. The amplifier is applied to compensate the decrease in intensity of a barcode signal because of the subtraction of the slightly enveloped line from the original barcode signal. The black level of the average signal is relatively stable and this average signal has a wide enough difference between black and gray levels. Therefore, a black code signal is obtained by comparing the average signal with a threshold V_{T1}.

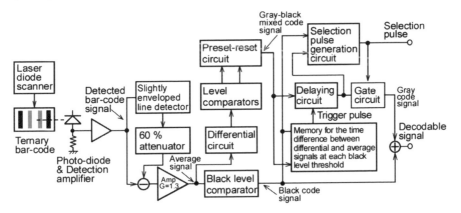

Figure 5.23. *Schematic outline of the TBDS employing the envelope differential composite method [WAK 08]*

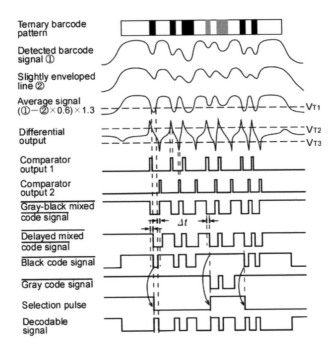

Figure 5.24. *Waveforms illustrating the operation principles of the envelope differential composite method*

This method automatically allows the system to maintain a level comparable to the appropriate one depending on the barcode signal.

When using high-density barcodes, a gray signal in the average signal is of small intensity and is not at a stable level. Therefore, to obtain a stable gray code signal, a differential method is applied to the average signal. The average signal is differentiated to extract the average signal waveform's transition, and the differential signal is compared with threshold voltages V_{T2} and V_{T3} employing two level comparators. Narrow comparator-output pulses corresponding to transitions from white to black or gray levels and the reverse are obtained. A gray–black mixed code signal is then obtained by a presetting operation at the preceding rising edge of these comparator-output pulses and a resetting operation at the following rising edge using a preset–reset circuit. Black and mixed code signals are obtained using the same processing circuits as those described in section 5.4.1. Because the system detects a ternary barcode with narrow bars while eliminating signal distortion using the slightly enveloped method and emphasizing the transitions of the barcode signal applying the differential method, the system can steadily detect high-density barcodes.

5.4.2.2. *Experimental results*

A prototype system for a single-line four character ternary barcode has been developed and tested. The detectable range for a low minimum bar width of 0.3 mm was found to be more than 5.4 cm, which is a wide range (Figure 5.25). This range is 4.5 times greater than that found with the DT method. This result means that correct detection is not mainly dependent on the detection distance, since approximately unbent averaged signals that adapt to any barcode pattern with little distortion are obtained over a wide range by applying the slightly enveloped and differential methods. As the minimum bar width W decreases, the detection signal for a narrow bar becomes small and so the detection range deteriorates. Contrarily, as W increases, the minimum detection length lengthens since the average signal distortion for wide black bars becomes remarkable near the minimum detection length because of the small $\tau(=7.3 \ \mu s)$ for the enveloped line detector. However, the detection range is almost the same as that used in the DT method, since the average signal distortion diminishes at distant detection lengths because of the short wide-black-bar signal rise time in the average signal. Concerning the scanning speed, the possibility of detection with a high scanning speed (351 scan/s) over seven times that of conventional CCD cameras was confirmed. Thus, the possibility of high-speed detection with a wide detection range was confirmed even when the density of BC was increased to a practical bar-width level of 0.3 mm.

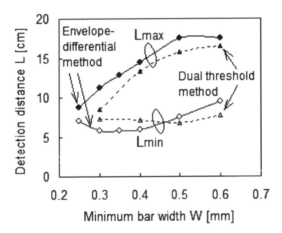

Figure 5.25. *Detection distance versus minimum bar width. The number of characters is 4 in addition to the start and stop codes. The scanning speed is 333 scan/s [WAK 08]*

5.4.3. *Fixed-period delay method*

The TBDS employing the envelope differential method has a count-latch memory for delaying the gray–black mixed code signals so as to separate them into gray and black code signals. Because of the distortion of the delayed mixed code signal through the counting period-latch timing instability and the fluctuation of the differentiated signal through the noise contained in the average signal, the detection bar width is limited to nearly 0.3 mm while a practical minimum bar width $W = 0.25$ mm is needed. A TBDS employing a fixed-period delay (FPD) method was developed in 2010 to provide a longer range and a higher scanning speed while being simple and able to handle much information [WAK 10b].

5.4.3.1. *TBDS with a fixed-period delay method*

The FPD TBDS using a laser diode scanner is outlined in Figure 5.26. The detection method is explained in Figure 5.27. In this system, the average signal is processed using a nonlinear filter (NLF) to suppress the sharp-edged noise contained in the average signal. The NLF output signal is differentiated to extract the NLF output signal's waveform. Comparator-output pulses corresponding to the transition from white to black or gray levels and the reverse are obtained by comparing the differential signal with the two thresholds V_{T2} and V_{T3}. A gray–black mixed code signal is obtained through a presetting operation at the preceding rising edge of these comparator-output pulses and a resetting operation at the following rising edge. The mixed code signal is then delayed for a fixed period t_d using a shift register to eliminate the delayed signal's distortion as in the count-latch method. Gray and adjusted black code signals and the decodable signal are obtained by the same operation as described in section 5.4.2. The system does not require complicated image processing and has a simple circuit configuration without causing an unstable delay. Therefore, a miniature TBDS with longer range and higher speed is expected to be realized.

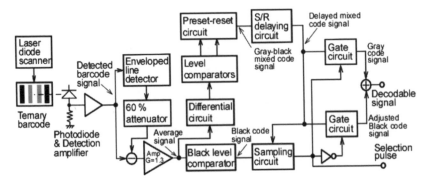

Figure 5.26. *Schematic outline of the fixed-period delay TBDS [WAK 10b]*

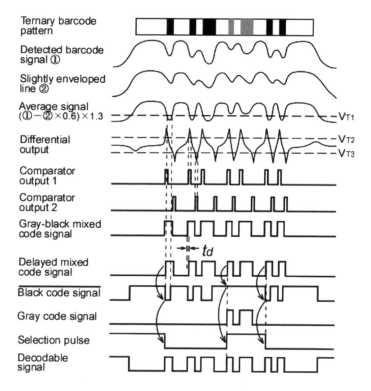

Detected waveform labels (left column): Ternary barcode pattern; Detected barcode signal ①; Slightly enveloped line ②; Average signal (①−②×0.6)×1.3; Differential output; Comparator output 1; Comparator output 2; Gray-black mixed code signal; Delayed mixed code signal; Black code signal; Gray code signal; Selection pulse; Decodable signal.

Figure 5.27. *Operation principle waveforms of the TBDS using the FPD method [WAK 10b]*

5.4.3.2. *Experimental results*

An NLF capable of cutting off high-frequency signals over 2.4 MHz and dropping sharply in gain and clipping the top and bottom sharp-edged signals over 100 kHz has been developed to suppress the sharp-edged noise (Figure 5.28). The unity gain frequency of the OP Amp used in this filter were ~3 MHz. The detection performance of the system using this NLF was examined. The detection distance versus minimum bar width of the system is shown in Figure 5.29, where it is compared with that of the count-latch system (without the NLF). The detectable range for the BC with $W = 0.25$ mm was increased to a practical range of 5.3 cm, which is 1.8 times longer than that in the count-latch system. The detection range was improved particularly for long distances, where the detection signals are easily distorted, employing a shift-register stable delay for the mixed code signal. Noise suppression was effective in detecting high-density barcodes. It also contributed to the detection range improvement of the BC with $W = 0.25$ mm; the improvement was about 1 cm.

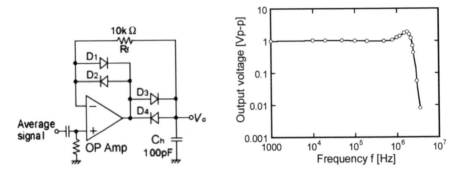

Figure 5.28. *(a) Configuration of a nonlinear filter. (b) Peak–peak output voltage versus frequency. The unity-gain frequency in the OP Amp f_T is ~3 MHz. The amplitude of the input sine wave signal is 0.5 V [WAK 10b]*

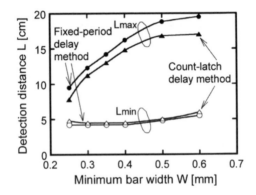

Figure 5.29. *Detection distance versus minimum bar width for the TBDS. $t_d = 2.5$ μs, $v_s = 333$ scan/s [WAK 10b]*

In examination of the possibility of high-speed detection, a maximum scanning speed of 417 scan/s, which is 8.3 times the ~50 scan/s achieved with the conventional CCD cameras, was obtained under the practical detection range for $W = 0.3$ mm. This is because there was no need for complicated image processing. A TBDS with a FPD method was shown to be capable of providing high-speed detection of a high-density barcode with $W = 0.25$ mm.

5.5. RFID

RFID is an identification technique that does not use light, and thus, it is essentially different from the barcode technologies. The basic idea of this technology was devised to identify aircraft in the World War II era [RFI 04, STE 05]. RFID employs a microwave transmission method using radio waves (950 MHz–2.45 GHz; UHF band) and an electromagnetic induction method using electromagnetic waves (13.56 MHz: shortwave band). At practical levels, radio wave technology and electromagnetic induction using magnetic waves of 2.45 GHz and 13.56 MHz, respectively, have been developed [RFI 04, JAP 03].

5.5.1. *Electromagnetic induction technology*

Electromagnetic induction technology using magnetic waves of 13.56 MHz is practical and is characterized by processing a large amount of information. The technology uses a reader/writer to emit electromagnetic waves and an IC tag with an antenna coil and memory to receive the wave and transmit saved information (Figure 5.30). For RFID employing the electromagnetic induction method, IC tags with loop coils are located at a maximum of 1 m from a reader/writer with a loop coil. 13.56 MHz transmitting signals are used to create an induction electromagnetic field. When the modulation signal with read and write instructions is transmitted from the coil of the reader/writer, a modulation signal is induced in the coils of the IC tags. Power is then generated and demodulation achieved on the IC tags. IC tags read data from memory and write data. When the IC tags transmit the modulation signal including read and write results to the reader/writer, the reader detects the modulation signal and demodulates it. ASK (Amplitude Shift Keying) for modulating a transmission signal amplitude, FSK (Frequency Shift Keying) for modulating its frequency, and PSK (Phase Shift Keying) for modulating its phase are employed for the modulation method. This technology is currently used for automatic ticket inspection systems, entering/leaving checking systems, auto matic checkout counter systems in cafeterias, and identification at libraries and of containers in airports. The amount of information is from several bytes to several kilobytes. The transmission speed is several kilobytes to 212 kb/s.

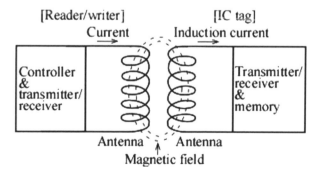

Figure 5.30. *Electromagnetic induction method*

The transmission directivity of the electromagnetic wave from a reader/writer is gradual, and so, this technology has wide communication ability. This technology also has features capable of writing information into IC tags and reading them in the presence of stains and dust. However, detection is affected by the installation direction of IC tags. This is difficult to use owing to a detection error problem (related to an eddy current effect) in the case of nearby conductive materials such as metals [JAP 03, YAM 05]. Another drawback of this RFID is that the cost of IC tags is high. If these several issues are resolved, the technology will be used in the wide area of FA, physical distribution, electronic ticketing, and security.

5.5.2. *Microwave transmission technology*

In microwave technology using magnetic waves of 2.45 GHz, a reader/writer transmits data to and receives data from IC tags (Figure 5.31). Antennas for a reader/writer and IC tags are microstrips printed on the circuit substrate, which is shorter than that in the electromagnetic induction system. Microwaves are modulated by transmission data and transmitted to IC tags. IC tags receive these data to demodulate and achieve reading and writing instructions. The linear polarization and circular polarization methods are used for radio waves. In the former method, the transmission and reception direction of radio waves must be adjusted to the oscillating direction of the radio waves. Therefore, there is a limitation placed on the installation of the reader/writer and IC tags. On the other hand, in the latter method, since the electric field direction of radio waves circulates in transmission, there is no limitation to the installation of reader/writer and IC tags.

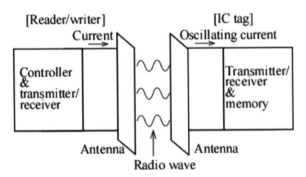

Figure 5.31. *Microwave transmission method*

Thus, this technology enables the reader/writer to communicate from several meters with the IC tags. It is appropriate to identify a large section of material. It is possible to have directivity for the reader/writer. It has the same features as described in section 5.5.1, capable of writing information into IC tags and reading them in the presence of stains and dust. It has also the feature of reading IC tags in rough positions. However, this is easily influenced by reflection from metal [YAM 05]. It is also influenced by water and humans through absorption, and its communication distance is thus shortened [JAP 03]. When IC tags of 2.45 GHz are used nearby radio-wave local area network (LAN) units, radio-wave communications are apt to disturb each other. This is the case in controlling the comings and goings of cars in a parking area and product control on assembly and processing lines (FA). In the future, the technology will be used in the package and material management of physical distribution in addition to these areas because it is possible to communicate from long distances.

5.6. Application examples

Serial binary barcode and the two-dimensional BCDSs will be typically used in the following areas because of low-speed identification processing capability.

– Low-speed sorting of parcels in automatic production lines (FA) and automated warehouses;

– Receipt and shipment of goods;

– Package and material management in physical distribution;

– Management of electronic and mechanical components;

– Security control;

– Management of medicine and blood;

– Library book management.

The multi-line-scan 2D binary barcode and the TBDS using a laser scanner will be used especially for high-speed sorting of parcels in automatic production lines and automated warehouses in addition to the areas listed above because of the high-speed identification and handling capability of large amounts of information.

RFID technology can be used in the following application areas because of wide communication capability.

– Electronic ticket inspection systems;

– Checkout counter systems in cafeterias;

– Identification at libraries;

– FA (product control in assembly and processing lines);

– Security control (e.g. entering/leaving checking system);

– Parking control;

– Package and material management in physical distribution;

– Management of livestock;

– Tracing of animal movement.

However, RFID technology is apt to be influenced by water, metal, and radio-wave networks. Therefore, it should be used in an environment without these materials or with the application of new techniques (such as applying an amorphous insulator) that eliminate the influences.

5.7. Concluding remarks

Current technologies and technologies in their research phase for automated identification were described, starting with the historical development of the binary barcode and radio-frequency identification technologies. The main current barcodes – serial binary barcodes and two-dimensional binary barcodes – and related detection technologies of laser scanners and CCD line scanners and CCD/MOS area solid image sensors were introduced. These technologies are currently fully grown and so are predicted to be combined with other technologies such as RFID or to be merged into multi-line scanning with time-sharing laser light emission operation and ternary barcode detection. Multi-line scanning technology employing the TSLE method and pulse modulation drive method of LDs for realizing effective high-speed scanning were also introduced in their research phase. As high-speed scanning technologies of barcodes containing large amounts of information, the ternary barcode and detection techniques of a DT method, an envelope differential composite method, and a fixed-period delay method in their research phases were then described. These research phase techniques are thought to be useful for applications in which high-speed identification is necessary, while investigation into optimal miniaturization of optical systems is needed. Furthermore, current RFID technologies and their future prospects were introduced. In terms of RFID, new technologies to provide solutions to reflection by metal and absorption by water, and interference between radio-wave LAN units and RFID units are hoped to be developed in the near future.

5.8. Acknowledgments

The author would like to thank President Y. Aragane and Dr. K. Watanabe of Tokyo Metropolitan College of Industrial Technology for their support concerning the publishing of this document. He would also like to thank Dr. C. Nagasawa of Tokyo Metropolitan University for his valuable support regarding works on multi-line barcode and ternary barcode detection systems. He would also like to thank Mr. O. Hayashiguchi and Mr. H. Ajiki for discussions throughout research on the multi-line barcode detection system. In addition, he would like to thank Dr. Nagasawa, IEEE, Elsevier, the Japan Society of Applied Physics, and Japan Industrial Publishing Co., Ltd., for their kind permission to reuse figures from previously published materials [HIR 03, WAK 00a, WAK 01, WAK 06a, WAK 08, WAK 10b].

A part of research concerning the multi-line barcode detection technology was supported by Grants-in-Aid for Scientific Research from the Japan Society for the Promotion of Science awarded in FYs 2002 and 2003. A part of work on the ternary barcode detection technology was also supported by the program of Collaboration of Innovative Seeds by the Japanese Science and Technology Agency in FY2009.

5.9. Bibliography

[HIR 01] HIRAMOTO J., *Knowledge of Barcode and Two-Dimensional Code*, Tokyo, Japan Industrial Publishing Co. Ltd., 2001.

[HIR 03] HIRAMOTO J., *Barcode Symbol (1D·2D) Introductory Text*, Tokyo, Japan Industrial Publishing Co. Ltd., 2003.

[JAP 03] JAPAN AUTOMATIC IDENTIFICATION SYSTEMS ASSOCIATION, *Radio Frequency Identification*, Tokyo, Ohmsha Co., 2003.

[LIY 95] LI Y., KATZ J., "Laser beam scanning by rotary mirrors. I. Modeling mirror-scanning devices", *Applied Optics*, vol. 34, no. 28, 1995, pp. 6403–6416.

[NIK 03] NIKKEI COMPUTER, IC tag (RFID), Tokyo, Nikkei BP, 2003.

[RFI 04] RFID TECHNOLOGY EDITORIAL DEPARTMENT, *All Radio-Frequency IC Tags*, Tokyo, Nikkei BP, 2004.

[SEQ 75] SEQUIN H.C., TOMPSETT F.M., *Charge Transfer Devices*, New York, Academic Press, Inc., 1975.

[STE 05] STELLUTO C.G., *The State of RFID Implementation and its Policy Implications*, NJ USA, IEEE-USA EBooks, 2005.

[TAD 03] TADOKORO Y., *Measurement and Sensor Engineering*, Tokyo, Ohmsha Co., 2003.

[WAK 96] WAKAUMI H., AJIKI H., "High-speed two-dimensional bar-code detection system", *The 1996 Conference on Lasers and Electro-Optics Europe, 1996 CLEO/Europe*, Hamburg, Germany, 8–13 September 1996, Technical Digest, p. 188.

[WAK 98a] WAKAUMI H., AJIKI H., "Two-dimensional bar-code detection system using a complementary laser light emission method", *Sensors and Materials*, vol. 10, no. 1, 1998, pp. 47–61.

[WAK 98b] WAKAUMI H., AJIKI H., "A high-speed 12-layer two-dimensional bar-code detection system with wide-band photo-detection amplifier and balanced raster scanner", *Proceedings of SPIE – The International Society for Optical Engineering*, vol. 3491, 1998, pp. 868–872.

[WAK 99] WAKAUMI H., NAGASAWA C., "A three-layer two-dimensional bar-code detection system with time-sharing laser light emission method", *The Transactions of the Institute of Electronics, Information and Communication Engineers C-I*, vol. J82-C-1, no. 11, 1999, pp. 650–651.

[WAK 00a] WAKAUMI H., "A high-speed 12-layer two-dimensional bar-code detection system", *Optical Review*, vol. 7, no. 1, 2000, pp. 66–72.

[WAK 00b] WAKAUMI H., NAGASAWA C., "A high-speed two-dimensional bar-code detection system with time-sharing laser light emission method", *Proceedings of SPIE – The International Society for Optical Engineering*, vol. 4087, 2000, pp. 1253–1258.

[WAK 01] WAKAUMI H., NAGASAWA C., "Development of a two-dimensional bar-code detection system using multi laser diodes with time-sharing light emission operation", *Optical Review*, vol. 8, no. 2, 2001, pp. 101–106.

[WAK 04] WAKAUMI H., NAGASAWA C., "High detection resolution for two-dimensional bar-code detection system using masked collimators", *Sensors and Actuators A: Physical*, vol. 110, no. 1–3, 2004, pp. 177–181.

[WAK 05] WAKAUMI H., NAGASAWA C., "A ternary bar-code detection system with pattern-adaptable dual threshold", *Eurosensors XII Extended Abstracts*, Barcelona, Spain, 11–14 September 2005, p. MP43.

[WAK 06a] WAKAUMI H., NAGASAWA C., "A ternary barcode detection system with a pattern-adaptable dual threshold", *Sensors and Actuators A: Physical*, vol. 130–131, 2006, pp. 176–183.

[WAK 06b] WAKAUMI H., NAGASAWA C., "A 2D ternary barcode detection system with a dual threshold", *Proceedings of the 5th IEEE Conference on Sensors, IEEE Sensors 2006*, Daegu, Korea, 22–25 October 2006, pp. 1511–1514.

[WAK 07] WAKAUMI H., NAGASAWA C., "A high-speed 2D barcode detection system with a dual threshold method", *International Conference on Control, Instrumentation and Mechatronics Engineering*, CIM'07, Johor Bahru, Malaysia, 28–29 May 2007, pp. 378–383.

[WAK 08] WAKAUMI H., "A high-density ternary barcode detection system employing an envelope-differential composite method", *Proceedings of the 7th IEEE Conference on Sensors*, IEEE Sensors 2008, Lecce, Italy, 26–29 October 2008, pp. 1076–1079.

[WAK 09] WAKAUMI H., "A high-speed six-line ternary barcode detection system with a dual threshold method", *ICROS-SICE International Joint Conference 2009*, Fukuoka, Japan, 18–21 August 2009, Final Papers, pp. 1127–1131.

[WAK 10a] WAKAUMI H., "A six-line ternary barcode detection system with a dual threshold method", *International Journal of Mechatronics and Manufacturing Systems*, vol. 3, no. 3–4, 2010, pp. 261–273.

[WAK 10b] WAKAUMI H., "A high-density ternary barcode detection system with a fixed-period delay method", *Eurosensors XXIV*, Linz, Austria, 5–8 September 2010, pp. 252–255.

[YAM 05] YAMANAKA T., KONDA J., URATA T., "The effect of eddy-current flux on transmission capacity of RFID antennas embedded in mobile phones", *IEICE Technical Report-EMCJ*, vol. 105, no. 439, 2005, pp. 1–6.

Chapter 6

An Active Orthosis for Gait Rehabilitation

Gait therapy is vital for restoring neuromuscular control in patients suffering from neurological injuries. Robots can provide prolonged, systematic, and repetitive gait training sessions. Currently available robotic devices use stiff actuators with high end point impedance. This work presents a new compliant robotic gait rehabilitation system. Pneumatic muscle actuators (PMAs) were used for actuation purposes. The robotic device is lightweight and works in perfect alignment with patient's joints. The modeling of robotic device with PMA was performed. Model reference-based adaptive control (MRAC) was used to guide the patient's limbs on physiological gait patterns, and joint torques required to achieve these trajectories were measured. The PMA with the proposed design is capable of providing the required joint torques. Simulation studies are reported.

6.1. Introduction

6.1.1. *Gait rehabilitation*

Neurologic injuries such as stroke and spinal cord injuries (SCI) cause damage to neural system and motor function, which results in lower limb impairment and gait disorders. Patients with gait disorders require specific training to regain functional mobility. Traditionally, manual physical therapy has been used for gait rehabilitation of neurologically impaired patients. Body weight–supported (BWS) manually-assisted treadmill training has been in practice for more than 20 years (Figure 6.1) [FIN 91, HES 95, BEH 00]. It allows the patient to perform a favorable gait for

Chapter written by Shahid HUSSAIN and Sheng Q. XIE.

greater balance training and longer stance durations compared with over ground gait training [MUR 85, HAS 97, HES 99, PAT 08]. BWS treadmill training has also proven significant improvements in step length, endurance, and walking speed of neurologically impaired patients [HES 95, VIS 98, LAU 01, TEI 01].

The quality of manually-assisted BWS treadmill training is dependent on the therapist's experience and judgment, which varies widely among the therapists. The BWS training requires a team of three therapists to train the patient's limbs and to stabilize the pelvis, which increases the cost of therapy. The training sessions are normally short due to the physical therapist's fatigue. Manually-assisted training also lacks proper methods of recording the patient's progress and recovery.

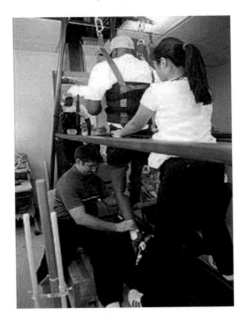

Figure 6.1. *Manually-assisted BWS treadmill training*

6.1.2. *Rehabilitation robotics*

Automated rehabilitation solutions have been researched lately to overcome the above-mentioned shortcomings of manually-assisted training [COL 00]. Robot-assisted gait training has several advantages over manually-assisted treadmill training. It relieves the physical therapist from the strenuous task of manual assistance and facilitates in delivering well-controlled repetitive and prolonged gait training sessions at a reasonable cost. The physical therapist's role is limited to supervision. The subjectivity of a manual training process is eliminated by providing measurement of interaction forces and limb movements to assess the quantitative level of improvement in gait parameters.

The history of robotic rehabilitation started with the adaptation of industrial robotic manipulators to the field of physical therapy [NAP 89, BOL 95, HOG 00]. Following that trend, various devices have been designed for restoration of upper limb and gait functions. The industrial robotic manipulators are mainly designed for tasks such as pick and place and are inherently stiff and massive. However, robotic rehabilitation devices need compliant and safe human–robot interface [VEN 06, SUG 07]. Subsequently, robots for applying suitable forces and capable of providing a safe interaction with the patients have been developed [COL 00]. Most of these robots are wearable and work in proximity with the patient's limbs. *Active orthosis* is a more common term for these wearable robotic devices. From the studies of human gait biomechanics and manual physical therapy practice, different gait training strategies are incorporated in the robot control schemes to enhance the rate of recovery.

The process for developing, testing, and analyzing the efficacy of robotic gait rehabilitation orthoses involves four stages (Figure 6.2). Stage 1 involves the process of determining kinematic and kinetic constraints for the design of active orthosis. Studies from the fields of clinical gait analysis and human gait biomechanics provide the basic criteria for the design of these robotic devices. Stage 2 is to design the active orthosis, which can be adjusted to patients with different anthropometric parameters. Stage 3 is to select a suitable gait training strategy according to the patient's disability level and phase of rehabilitation. Stage 4 is to evaluate the functional outcomes of robot-assisted gait rehabilitation and adjust the gait training parameters accordingly. The functional outcomes involve the improvement in gait parameters like stride length, stepping frequency, stance duration, and muscle coordination patterns.

The main focus of this work was the development of a compliant active gait training orthosis and a gait training strategy based on adaptive control. The proposed orthosis provides a compliant and safe interaction with the patient, and the adaptive gait training strategy enhances the patient's voluntary participation in the gait training process. The patient's interaction with the active orthosis was estimated using a combined patient-active orthosis dynamic model. Section 6.1 follows with an overview of the biomechanics of human gait to familiarize the reader with the relevant concepts used in the development of active orthosis. A review of existing active gait rehabilitation orthoses and gait training strategies is also provided. Section 6.2 deals with the design of compliant active orthosis. Sections 6.3 and 6.4 present the modeling and controller design of compliant active orthosis, respectively. Section 6.5 presents the simulation results of the proposed orthosis controller. Section 6.6 contains conclusions.

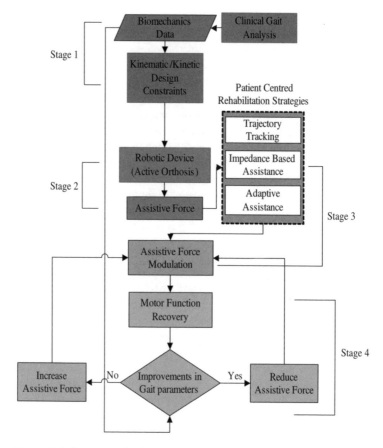

Figure 6.2. *Overview of robotic gait rehabilitation process: Stage 1, Stage 2, Stage 3, and Stage 4*

6.1.3. *Biomechanics of gait*

A background of gait biomechanics is provided in this section to familiarize the reader with the concepts used in the design of active gait training orthoses and training strategies. During the past 50 years, there have been major advancements in the field of biomechanics particularly associated with kinematic and kinetic analysis of human gait [WIN 91, AND 03, SHE 06, SET 07]. Comprehensive knowledge of physiological gait patterns is now available, which has facilitated researchers to design improved robotic orthoses and training strategies for effective motor function recovery. The knowledge of gait biomechanics is also important to determine the efficacy of robot-assisted gait training in gait analysis laboratories [KAO 10, MUL 10].

A *gait cycle* [WIN 90] (Figure 6.3) is the sequence of events from the heel strike of one foot to the subsequent heel strike of the same foot [WIN 90, WIN 91]. It is defined in terms of time interval and usually expressed as a percentage of gait events taking place. Walking consists of repeated gait cycles [WIN 90]. The gait cycle consists of two phases: stance and swing. The *stance phase* is defined by the percentage of gait cycle when the foot is in contact with the ground and the *swing phase* by the time when the foot is in air and not bearing any load. Approximately 62% of the gait cycle consists of stance and 38% of swing phase.

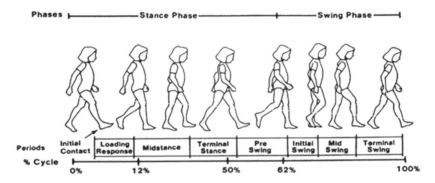

Figure 6.3. *Phases of gait cycle*

Human gait is realized by coordinated inter joint movements of the lower limb. Muscles are used to create moments across these joints. Three planes divide the human body into six parts (Figure 6.4a). The hip joint can provide motions in all the three planes as hip flexion/extension in the sagittal plane, abduction/adduction in the frontal plane, and rotation in the transverse plane. The knee joint has major rotations in the sagittal plane as flexion/extension and also provides rotations in the transverse plane. The ankle is a complex joint and due to its variable center of rotation, the axes of motion are not simply three Euclidean axes. The important one is plantar/dorsiflexion in the sagittal plane for ground clearance during the swing phase. The sagittal plane joint ranges of motion (Figure 6.4b) and moments contribute most during the gait cycle and are actuated in most of the active gait training orthoses. Gait biomechanics and analysis is an important research area for analyzing the outcomes of robot-assisted gait training [ZIS 07, ALI 09]. The standard procedure involves kinematic data collection (joint angles, velocities, and accelerations) by using reflective markers and motion capture systems, whereas kinetic data (joint moments and power) is obtained by measuring ground reaction forces with the aid of foot plates. Electromyography (EMG) signals are used to judge the activity of various muscle groups in combination with kinematic and kinetic data.

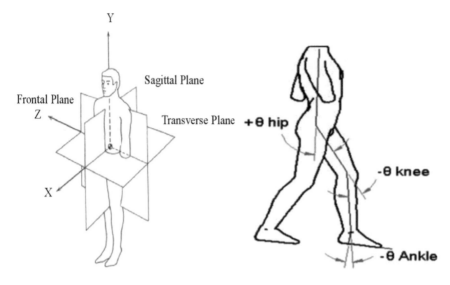

Figure 6.4. *(a) Plane and axis [WIN 91, WIN 90]; (b) joint angles in the sagittal plane [WIN 91]*

6.1.4. *Robot-assisted gait rehabilitation: a review*

Active orthoses are training devices that work in parallel with the human body and have mechanical actuation to apply forces to the human limbs. The history of active orthoses started in late 1970s. Early active orthoses were standard braces with added actuation mechanisms [VUK 74]. Among the first full lower limb active orthosis is the University of Wisconsin prototype [SEI 81]. The orthosis has universal joints at the hip and ankle and provides sagittal plane flexion/extension motions by means of hydraulic cylinders. The remaining degrees-of-freedom (DOFs) are passively held by springs.

The first modern automated BWS treadmill training system Lokomat was developed in the late 1990s and is commercially available. The system has a wearable driven gait orthosis (DGO) having mechanical actuation to power hip and knee sagittal plane rotations [COL 00]. Direct current (DC) motors with a ball screw mechanism are used to power these joints. Dorsiflexion to the ankle joint is provided by passive elastic bands, and the hip abduction/adduction is kept free. DGO works on the assumption that the orthosis joints are in perfect alignment with the patient's joints, and the joint positions are measured with encoders built in DC motors. The physical contact between the DGO and the patient is through two force-torque sensors placed in series with DC motors that move orthosis links. DGO is connected to the treadmill by a rotatable parallelogram linkage to stabilize the patient's trunk.

In this manner, the DGO moves only in a vertical direction, avoiding any sideways tilt of trunk. Active leg exoskeleton (ALEX) was developed at the University of Delaware for gait training of stroke survivors [BAN 09]. ALEX uses gravity-balancing orthosis (GBO) [AGR 04] as its foundation. GBO is a passive device without any mechanical actuation and uses the conventional method of fixing the center of mass by means of a spring mechanism. Linear servo drives have been used on the GBO for providing actuation at the hip and knee joints for flexion/extension rotations in the sagittal plane. Hip abduction/adduction and four trunk rotations are held passive by means of springs.

Lower extremity powered exoskeleton (LOPES) uses a Bowden cable-based actuation system [VEN 07]. It is built on the idea of a lightweight exoskeleton system having a pair of springs in series with an electric motor. The electric motor is coupled to the springs via Bowden cables. Due to the cable-based actuation, the electric motor is placed on a remote station and acts as a low weight pure force source. The displacement of springs recorded by linear potentiometers is used as a force measurement. The actuated DOFs include two pelvis rotations: hip sagittal and frontal plane rotations and knee sagittal plane rotations.

Ambulation-assisting robotic tool for human rehabilitation (ARTHUR) was developed to mechanically interact with a single leg during treadmill training. It consists of two moving coil brushless servo motors that drive either end of a two bar linkage [EMK 06]. ARTHUR provides motions to the knee and ankle joints in the sagittal plane. Pelvic Assist Manipulator (PAM) is being developed to allow naturalistic movements to the human pelvis during gait training [AOY 07]. Two 3-DOF robotic arms are used to assist the patient's pelvis during treadmill training. These two robotic arms are placed at an angle to give the therapist access from the side and from behind. PAM uses pneumatic actuators to provide lateral and rotational pelvic movements for the patient. PAM is used in combination with Pneumatically Operated Gait Orthosis (POGO), a device that provides pneumatic actuation for hip and knee sagittal rotations.

Although the robotic orthoses can provide systematic and prolonged treadmill training sessions, there are some drawbacks associated with their designs. Two approaches are seen in actuator placement for powering the active orthoses. In one approach, the actuators are placed on a remote station, and the actuation is transferred to the orthosis via cables, rigid linkages, and pneumatic or hydraulic systems [AOY 07, VEN 07]. The benefit of this method is that there are no limitations on actuator weight and hence the power capacity of the actuators. Inefficient transfer of power, non-durability of actuation transfer mechanism (cables), and lack of precise control are the drawbacks associated with this approach. In another approach, the actuators are directly mounted on the orthosis frame [COL 00, BAN 09]. The main advantage of this approach is the efficient transfer of power and a good alignment of orthosis joints with patient joints. The weight of actuators and gear assembly increases the overall weight of the orthosis. Reduction in weight of actuation mechanism reduces the maximum moments that could be applied to the patient's joints. Gravity balancing techniques have been developed to compensate for the weight of the orthosis by using spring and counter weight mechanisms [AGR 04].

6.1.5. *Gait training strategies: a review*

The goal of robot-assisted gait training is to reinstate neuroplasticity so that the motor function could be improved. Although successful determinants of gait training are largely unknown, repetitive and task-oriented training strategies may result in significant improvements [BAY 05, PAT 07]. These determinants have been formulated by drawing concepts from rehabilitation, neuroscience, and motor learning literature [KWA 97, BAR 06]. Gait training is to be provided according to the level of disability while encouraging the patient's active participation in the training process. Robot-assisted treadmill training utilizes trajectory tracking, impedance, and adaptive control-based training strategies.

Trajectory tracking or position control is widely implemented by robotic training devices. Trajectory tracking works on the principle of guiding the patient's limbs on fixed reference gait trajectories. It mainly consists of proportional feedback position controllers with joint angle gait trajectories as input [LUM 93, LUM 95, LUM 02]. For trajectory tracking, the issue of determining the reference trajectory is important. Mathematical models of normative gait trajectories and pre-recorded trajectories from healthy individuals are commonly used. A *teach and replay* technique has been introduced by the designers of ARTHUR in which a joint angle trajectory is recorded during manual assistance and is then replayed during robotic assistance [EMK 08]. Recently, a reference trajectory generation method has been developed for hemiparetic patients. The desired trajectory for the impaired limb is generated online based on the movement of unimpaired contralateral limb [VAL 09].

Trajectory tracking is suitable for training patients with SCI or acute stroke when they have no muscular strength to move their limbs. A limiting feature of trajectory tracking is the imposition of a predefined trajectory, leaving the system inflexible to considering the patient's intention and capabilities. For patients having some muscular strength, trajectory tracking may cause damage to their neuromuscular system when they try to resist the fixed forces applied by actuators [LUM 06, PAT 06]. This may result in abnormal gait pattern generations and would leave the patient unable to adapt to physiological gait [KAH 06].

The patient's active participation and involvement in the robotic gait training process is important to develop neuroplasticity and motor control [EMK 05, KAE 05].

The terms *patient cooperative, assist as needed, compliant,* and *interactive robot-assisted gait training* are used in the literature [RIE 05, VEN 07, WOL 08]. Robot-assisted gait training uses impedance and adaptation-based control strategies to actively involve the patient in the training process.

The relationship between the force exerted by the actuators and the resulting motion is generally known as *Mechanical Impedance.* The concept of impedance control in the field of robotics is first introduced by Hogan [HOG 85]. The impedance controller works on the principle of *force-based impedance control* and is mostly implemented in the form of an outer position feedback loop and inner force feedback loop. Lokomat also uses an impedance controller of the same form [RIE 05]. For gait training purposes, the idea behind impedance control is to allow variable deviation from reference gait trajectory depending on the patient's resistance. As long as the patient is on the reference trajectory with minimum deviations, the robot should not intervene. After a set limit is exceeded, an adjustable moment is applied at each joint to keep the leg within a defined range along the reference trajectory. For higher impedance values, the concept of admittance control is also used by Lokomat [HOO 02]. An admittance controller as opposed to impedance control works on the principle of position-based impedance control. More recent forms of impedance controllers use the concept of viscous force fields [COL 05, CAI 06]. For ALEX, a force field controller is developed for applying tangential and normal forces at patient's ankle. The linear actuators mounted at the hip and knee joints simulate the forces applied at the ankle. Tangential forces help to move the patient along the trajectory, and normal forces simulate virtual walls around the desired ankle trajectory in the plane containing human thigh and shank [HOG 06, BAN 09]. LOPES also uses impedance control for its "patient in charge" and "robot in charge" modes [VEN 07]. For the robot in charge mode, the controller stiffness is increased, so the patient is not in a compliant environment.

The potential issue with trajectory tracking and impedance control-based training is that they do not tune controller parameters based on real-time judgment of the patient's abilities. Adaptive assistance is used to enhance the patient's active participation in the training process [EMK 07]. The basis of adaptive assistance is to modify the robot motion in a way that is desired by the patient.

Adaptive assistance is used for real-time tuning of the controllers designed for stiff robotic actuators to match the patient's disability level and to actively involve him or her in the training process. In adaptive assistance mode, robot motion is initiated from the physical interaction between the patient and the orthosis. As the disability level varies from subject to subject, online estimation of patient–orthosis interaction force is the most crucial task in the adaptive assistance paradigm. In most of the gait training orthoses, this interaction force is estimated from the combined patient–orthosis dynamic model. Different methods are used to estimate the patient–orthosis interaction torque component.

Lokomat uses a moving average-based exponential forgetting technique for interaction torque estimation. After obtaining this estimate, various joint angle adaptation algorithms are formulated to adapt reference gait trajectory parameters by online optimization. These algorithms include inverse dynamics-based joint angle adaptation, direct dynamics-based joint angle adaptation, and impedance control-based joint angle adaptation [JEZ 04]. Later an impedance magnitude adaptation algorithm was formulated for Lokomat [RIE 05]. This algorithm works based on the impedance magnitude adaptation with constant reference joint angle trajectories. When a smaller patient resistance is estimated, the controller impedance is set high to guide the patient's limbs on reference trajectory. Impedance magnitude is reduced in larger estimates, and larger deviations from the reference trajectory are allowed. ARTHUR uses a manual teaching approach of and a replay for robot-assisted gait training. Physical therapists are asked to impart manual gait training to the subjects first, and the kinematic and kinetic gait parameters are recorded. These recorded parameters are then used during robotic gait training to adapt the stiffness and damping of a proportional and derivative (PD) force controller as a function of trajectory tracking error [EMK 08].

The adaptive algorithms discussed above estimate the patient–orthosis interaction force from the combined dynamic model of the patient and orthosis mechanism. The quality of interaction force estimation depends on the accuracy of force and joint position sensors and also on the estimation algorithm [ERD 10]. The abrupt forces like muscle spasms arising from patient and resulting actuator non-backdrivability present a major problem to the interaction torque estimation. *Backdrivability or compliance* is the ability of the robot being moved by the patient with low mechanical impedance to allow the patient's voluntary movements [CAM 09].

6.2. Compliant active orthosis design

The robotic orthoses discussed above are driven by electric motors attached to gear boxes which are highly stiffened and supply very large torques in repose to the patient's clonus and strong spasms. This may result in injury to a patient. Electric motors thus present a mismatch in the compliance of the actuator and limb being assisted. Impedance and adaptive control has had success in addressing this problem but adds another layer of complexity and extra cost. The active orthosis design presented in this study is compliant because the active orthosis behaves softly and gently and reacts to the patient's muscular effort.

6.2.1. *Design criteria*

The first goal was to design an active orthosis for gait rehabilitation. For design purposes, the biomechanics of human gait was studied. To meet the functional and structural requirements of the active orthosis, the orthosis joints should work in perfect alignment with the patient's joints. Also the active orthosis should inhibit excessive knee and hip extension. The actuation system should be powerful enough to guide the patient's limbs on reference trajectories and be able to produce required joint moments (Table 6.1). Actuators should be highly back-drivable with low mechanical impedance to accommodate abrupt forces arising from clonus. The active orthosis must have safety limits at the ends of maximum ranges of joint motion, and the orthosis should return to an anatomical standing position if the actuation mechanism fails. Regarding the cosmetic requirements and ease of use, the active orthosis should be lightweight, easy to wear, and comfortable. The orthosis should also allow fast adjustment to individual patients with different anthropometric parameters. The actuation system should generate no perceivable noise.

Degree-of-freedom	Range of motion	Joint moment
Hip flexion/extension	$+60°/-30°$	55 Nm
Hip abduction/adduction	$+15°/-15°$	25 Nm
Knee flexion/extension	$+0°/-90°$	55 Nm

Table 6.1. *Joint ranges of motion and moments*

6.2.2. *Active orthosis components*

Actuated and free DOFs for the active orthosis were decided based on the joint ranges of motion. The major rotations during gait cycle are in the sagittal plane (Table 6.1). The actuated DOFs were hip and knee sagittal plane rotations. Besides the sagittal plane, hip abduction/adduction provides the second-largest motion. This DOF was kept free. A new type of compliant actuator, PMA (Figure 6.5), was used for providing actuation to active orthosis. Although the design with PMA was a difficult task, as they can provide only unidirectional pulling force, the optimal DOFs necessary to provide physiological gait pattern were chosen. To provide actuation at hip and knee joints, various mechanisms were studied to transfer the actuation from PMA to the orthosis joints. An antagonistic disc-PMA mechanism was selected for actuation purposes. Double groove disks were used at hip and knee joints for sagittal plane motions. The orthosis frame was made from aluminum rectangular tubing to meet the strength requirements for torque transmissions (Figure 6.6). All the orthosis sections were made telescopic so that they could match the anthropomorphic features of a larger patient population.

Figure 6.5. *Pneumatic muscle actuator*

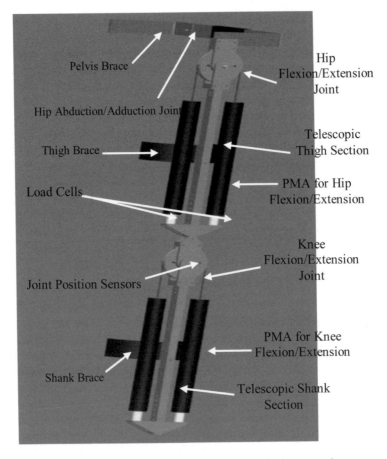

Figure 6.6. *Computer-aided design (CAD) model of active orthosis*

There were two reasons for using PMA for actuation purposes. The first reason is related to the geometrical design of active orthosis. The PMA has a high power to weight ratio, which makes it suitable for the task. The design is made simple and wearable, and the patient's joints will be in perfect alignment with the orthosis joints. The second reason is to introduce intrinsic compliance and back-drivability in the orthosis design. This compliance is beneficial for human–orthosis interaction and provides greater shock tolerance on heel strike, low actuator impedance, and more stable force control.

Absolute joint encoders were used at hip knee and joints to measure the angular positions. Load cells were used in series with each PMA to measure the pulling force generated. The ankle joint was not actuated.

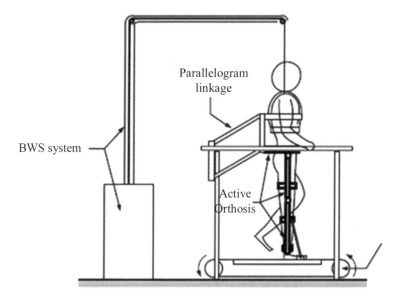

Figure 6.7. *Robot-assisted gait training*

The reason to omit an actuated ankle joint was that the ankle is a complex joint, and axes of motion are not simply three Euclidean axes. It is not necessary to provide an external ankle push off, as it could be done significantly with the aid of a treadmill. Also it is painful to apply an active force at the ankle joint without an individually fit-to-size foot interface. For safe and effective gait training, only ankle plantar/dorsiflexion is necessary for foot clearance during the swing phase. This dorsiflexion could be provided with the aid of some passive mechanisms like elastic straps or springs. Devices like pneumatically-driven ankle orthosis [FER 05] or Anklebot [ROY 09] can be added to the active orthosis if ankle actuation appears to be crucial from a clinical point of view. The schematic sketch of the complete system is shown Figure 6.7. A parallelogram linkage connects the active orthosis with the treadmill and also stabilizes the patient's pelvis in the vertical direction during training. A BWS system compensates for the weight of the patient and helps in foot clearance during the swing phase of gait.

6.3. Modeling

6.3.1. *PMA dynamic modeling*

Modeling of the active orthosis with PMA was a crucial task as they show highly nonlinear force-length characteristics. For this study, we considered the PMA model developed by Reynolds *et al.* [REY 03]. The modeled PMA has been inflated by supplying voltage to a solenoid that controls the flow of pressurized gas into the rubber bladder. It has been deflated by another exciting solenoid venting the contents of the bladder to the atmosphere. When inflated, the PMA shortens via the actions of the braided sheath, exerting a contractile force that is quite large in proportion to the PMAs weight.

The dynamic behavior of the PMA hanging vertically actuating a mass M has been modeled as a combination of a nonlinear friction, a nonlinear spring, and a nonlinear contractile element. The equation describing the dynamics of this PMA hanging vertically actuating a mass is:

$$M\ddot{x} + B(P)\dot{x} + K(P)x = F(P) - Mg \qquad [6.1]$$

where x is the amount of PMA contraction and the coefficients $K(P)$, $B(P)$, and $F(P)$ are given in [REY 03] as:

$$
\begin{aligned}
K(P) &= K_0 + K_1 P \\
&= 5.71 + 0.0307P
\end{aligned} \qquad [6.2]
$$

$$
\begin{aligned}
B(P) &= B_{0i} + B_{1i} P \\
&= 1.01 + 0.00691P
\end{aligned} \qquad [6.3]
$$

$$
\begin{aligned}
B(P) &= B_{0d} + B_{1d} P \\
&= 0.6 - 0.000803P
\end{aligned} \qquad [6.4]
$$

$$
\begin{aligned}
F(P) &= F_0 + F_1 P \\
&= 179.2 + 1.39P
\end{aligned} \qquad [6.5]
$$

From equation [6.1] the total force exerted by the PMA on the mass is:

$$\alpha = F(P) - B(P)\dot{x} - K(P)x \qquad [6.6]$$

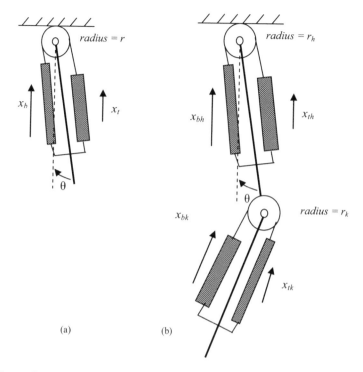

Figure 6.8. *(a) Antagonistic PMA configuration; (b) active orthosis with PMAs*

For the antagonistic configuration of PMA, the torque imparted to the joint by the PMA pair is (Figure 6.8a):

$$T_{total} = T_b - T_t = (\alpha_b - \alpha_t)r \qquad [6.7]$$

where T_b and T_t are the torque due to each of the individual PMA and are given by:

$$T_b = (F_b - K_b x_b - B_b \dot{x}_b)r \qquad [6.8]$$

$$T_t = (F_t - K_t x_t - B_t \dot{x}_t) \qquad [6.9]$$

where x_b is the length of PMA b, x_t is the length of PMA t, and r is the radius of disc. Thus, the relation for total torque (T_{total}) becomes:

$$T_{total} = \left(F_b - K_b x_b - B_b \dot{x}_b - F_t + K_t x_t + B_t \dot{x}_t \right)r \qquad [6.10]$$

The arrangement of the PMA on the active orthosis is shown in Figure 6.8b. Under these conditions, the hip and knee sagittal plane torques T_h and T_k, respectively, can be expressed using equation [6.10]

$$T_h = \left(F_h - K_h x_{th} - B_{th} \dot{x}_{th} - F_h + K_h x_{bh} + B_{bh} \dot{x}_{bh} \right) r_h \qquad [6.11]$$

$$T_k = \left(F_k - K_k x_{tk} - B_{tk} \dot{x}_{tk} - F_k + K_k x_{bk} + B_{bk} \dot{x}_{bk} \right) r_k \qquad [6.12]$$

where the subscripts h and k represent the coefficients for hip and knee joints, respectively.

6.3.2. *Interaction force estimation*

The estimation of patient-active orthosis interaction force requires a combined dynamic model of the patient and the active orthosis. As the anthropometric parameters and disability level vary from subject to subject, online estimation of patient–orthosis interaction force is the most crucial task in adaptive assistance paradigm. A robot dynamic equation was used for estimating the patient–orthosis interaction forces:

$$M(x)\overline{x} + C(x,\dot{x})\dot{x} + G(x) = T_a + T_p - T_f \qquad [6.13]$$

where M is the combined patient–orthosis inertia matrix, C is the combined patient–orthosis coriolis and centrifugal torque, G is a term representing gravitational torques, and T_f is the joint friction torque. T_a is the torque applied by actuator onto the orthosis and is measured by force sensors. T_p is the patient–orthosis interaction torque or the resistance offered by patient to applied actuator forces, and x is the generalized position vector representing joint angles. $\dot{x}$ and $\overline{x}$ are joint velocity and acceleration, respectively.

The real-time update of this patient resistance component is a challenging task, and the least squares method with exponential forgetting is used to update this component during the training process. Least squares with exponential forgetting is a useful method of dealing with variable patient-active orthosis interaction torque. The intuitive motivation is that the past data is generated by past parameters and should be discontinued for the estimation of current parameters. For more details, refer to [SLO 91].

6.4. Control

The benefits of robot-assisted rehabilitation might be increased by using more advanced robotic systems. Although different robot-assisted gait training strategies are discussed above, it remains to be demonstrated which is the most effective. One way to enhance motor function recovery is to develop a robot control algorithm that seamlessly optimizes the interaction between the active orthosis and the patient to provide as much therapeutic benefit as possible. To promote patient involvement in the rehabilitation process, we hypothesize that an MRAC will be suitable.

The controller estimates the patient–orthosis interaction in real time and modulates the actuator forces accordingly. The MRAC scheme will be helpful to accommodate patients with variable disability levels.

The issue of the justification of the MRAC naturally arises, given the variety of control types available. For the robot-assisted gait rehabilitation where the system dynamics are time and position dependent and where a substantial uncertainty in the system characteristics is produced by the unknown patient resistance torque properties, the model reference approach seems particularly well suited [DUB 79]. The reference model chosen at the discretion of the designer provides a flexible means of specifying the desired closed loop performance characteristics. The use of a model-based controller allows impedance and assistance to be controlled separately so that the orthosis can simultaneously be highly compliant and be able to provide enough assistive force to complete desired spatial movements.

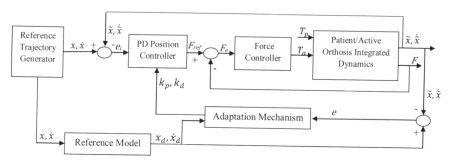

Figure 6.9. *Controller diagram*

The first step in the formation of MRAC is the selection of a reference model (Figure 6.9). The adaptation mechanism is designed to provide PD gains to minimize the trajectory tracking error. The PD position controller works on the principle of impedance control [HOG 85] and generates the reference force based on the trajectory tracking error. The force at the output is measured using a force-sensing system and is fed back to create a resulting error which is processed through a force controller to create torques applied at orthosis joints. The reference model chosen in this case is a linear, second-order, time invariant differential equation:

$$\overline{x}_d + 2\zeta\omega_n \dot{x}_d + \omega_n^2 = \omega_n^2 r(t)$$ [6.14]

Rewriting this expression:

$$a\overline{x}_d + b\dot{x}_d + x_d = r(t)$$ [6.15]

where ζ is the damping ratio, ω_n is the natural frequency, and a and b are given as:

$$a = \frac{1}{\omega_n^2} \text{ and } b = \frac{2\zeta}{\omega_n}$$

For the development of the algorithm, the coupling of the system DOFs is neglected, and the nonlinear manipulator dynamics are written assuming:

$$\bar{x} = \frac{ek_m}{m} \qquad [6.16]$$

where k_m is the actuator torque constant and m is the varying effective mass. The orthosis nonlinear dynamic equation can be written as:

$$\frac{m(i)}{k_m k_p}\bar{x} + \frac{k_d}{k_p}\dot{x} + \check{x} = r(t) \qquad [6.17]$$

The equation is of the form:

$$\beta(t)\bar{x} + \gamma\dot{x} + \check{x} = r(t) \qquad [6.18]$$

The adaptation mechanism was designed to provide PD gains to minimize the trajectory tracking error. The PD position controller works on the principle of impedance control [HOG 85] and generates the reference force based on the trajectory tracking error. In this study, it was unnecessary to obtain explicit knowledge of the coefficients β and γ. The force at the output was measured using force-sensing system and was fed back to create a resulting error which was processed through a force controller to create torques applied at orthosis joints. The patient will be able to train in a compliant and comfortable environment compared with the fixed trajectories applied by the existing gait rehabilitation orthoses. $x, \dot{x}$ are reference joint angle and velocity, respectively, $x_d, \dot{x}_d$ are joint position and velocity outputs of the reference model, $\bar{x}, \dot{x}$ are joint variables at output measured by joint sensors, e is the error fed to adaptation mechanism, and e_i is the position error fed to position controller. K_p and K_d are proportional and derivative gain values. F_{ref} is the reference force generated by PD controller; F_e is the error between the reference force and the force measured by force sensors (F).

A quadratic function is written in terms of difference between the responses and the actual referenced system as:

$$f(\varepsilon) = \frac{1}{2(q_0\varepsilon + q_1\dot{\varepsilon} + q_2\ddot{\varepsilon})} \qquad [6.19]$$

$$\dot{\beta} = -\frac{\partial f(\varepsilon)}{\partial \beta} = \frac{\partial f(\varepsilon)}{\partial a} \qquad\qquad [6.20]$$

$$\dot{\gamma} = -\frac{\partial f(\varepsilon)}{\partial \gamma} = \frac{\partial f(\varepsilon)}{\partial b} \qquad\qquad [6.21]$$

where ε is defined as $x_\mathrm{d} - \breve{x}$. After algebraic manipulation, the rates of adjustment of β and γ are:

$$\dot{\beta} = (q_0\varepsilon + q_1\dot{\varepsilon} + q_2\overline{\varepsilon})(q_0 u + q_1\dot{u} + q_2\overline{u}) \qquad\qquad [6.22]$$

$$\dot{\gamma} = (q_0\varepsilon + q_1\dot{\varepsilon} + q_2\overline{\varepsilon})(q_0 w + q_1\dot{w} + q_2\overline{w}) \qquad\qquad [6.23]$$

where the values of u and w and their derivatives are obtained from the solutions of the following differential equations:

$$a\overline{u} + b\dot{u} + u = -\overline{x}d \qquad\qquad [6.24]$$

$$a\overline{w} + b\dot{w} + w = -\dot{x}_\mathrm{d} \qquad\qquad [6.25]$$

After determining the values of γ and β, the rates of adjustment of the feedback gains of the system can be calculated by differentiating the definitions of γ (t) and β(t) obtained by comparing equations [6.17] and [6.18]. M is assumed to change slowly compared to the adaptation mechanism and during adaptive tracking β can be approximated by a. The result is:

$$\dot{k}_\mathrm{p} = -\frac{\dot{\beta} k_\mathrm{p}}{a} \qquad\qquad [6.26]$$

$$\dot{k}_\mathrm{d} = k_\mathrm{d}\dot{\gamma} - \frac{k_\mathrm{d}\dot{\beta}}{a} \qquad\qquad [6.27]$$

Explicit expressions for M are not required by the algorithm.

6.5. Simulation results

The active orthosis prototype leg with an antagonistic pair of PMA, actuating the hip and knee sagittal plane joints, is shown in Figure 6.8b. The simulation was performed using a fourth-order Runge–Kutta algorithm with a step size of 0.01 s.

The duration of the simulation was set to complete five gait cycles. The nominal joint angle trajectories of natural gait are reported in literature [WIN 91] and were used as reference trajectories for guiding the patient's limbs. The subject was considered to be completely passive, offering no resistance to the actuator torques. First stride from initial standing posture was eliminated, and the remaining four gait cycles were used for analysis purpose. Data from the right gait cycle was formatted for presentation purpose. The simulation was performed for a subject having a mass of 76.8 kg. Lengths of shank and thigh segment were 0.44 m and 0.43 m, respectively. The weight of the active orthosis (9.8 kg) was added to the subject's weight to form a mass matrix, M. A BWS of 40% was used.

The cadence was 85 steps/min. The reference joint angle trajectories were tracked with a maximum error of 1° and are shown as a percentage of gait cycle in Figure 6.10. The maximum hip angle achieved during flexion and extension was 16° and 14°, respectively. A knee flexion of 60° was achieved during mid-swing period. The torques required to track these trajectories as a percentage of gait cycle are shown in Figure 6.11. A peak hip torque of 50 Nm and knee torque of 90 Nm was achieved. The PMA can provide a peak force of 700 N satisfying the desired ranges of peak joint torques. The extent of actuator force adaptation also needs to be determined so that the gait patterns remain purely physiological. The interaction force feedback also needs to be evaluated as it may contain abrupt forces arising from patient's abnormal muscle functions like clonus or spasms.

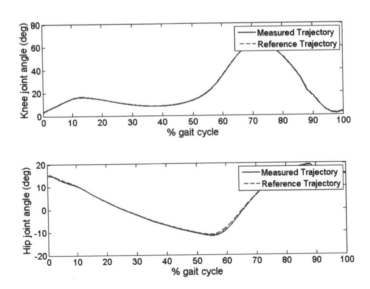

Figure 6.10. *Hip and knee joint angle trajectories as a percentage of gait cycle*

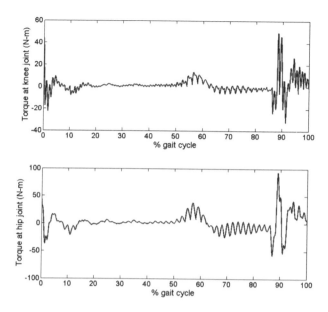

Figure 6.11. *Torque at hip and knee joints as a percentage of gait cycle*

6.6. Conclusions

Robotic orthosis can provide repetitive, prolonged, and systematic gait training sessions. A compliant gait rehabilitation orthosis has been designed. The variable level of disability among different patients represents a problem for devising a suitable gait training strategy. Also the nonlinear actuator dynamics make the control system design difficult. To overcome these problems, a dynamic model of the PMA actuators for the actuated DOFs was developed and was found to be suitable for the desired ranges of operation. To accommodate and train patients with variable disability, an MRAC algorithm was designed. The algorithm estimates patient-active orthosis interaction forces and adjusts the applied actuator forces accordingly. The performance of the MRAC was tested in computer simulations and was capable of guiding the patient's limbs on the physiological gait patterns. Future work involves the evaluation of the dynamic PMA model and MRAC on active orthosis prototype followed by clinical evaluations on the healthy and neurologically impaired subjects.

6.7. Acknowledgment

The authors would like to acknowledge the support of the Faculty Research Development Fund from the Faculty of Engineering, The University of Auckland, New Zealand.

6.8. Bibliography

[AGR 04] AGRAWAL S.K., FATTAH A., "Theory and design of an orthotic device for full or partial gravity-balancing of a human leg during motion", *IEEE Transactions on Neural Systems and Rehabilitation Engineering*, vol. 12, no. 2, 2004, pp. 157–165.

[ALI 09] ALIMUSAJ M., FRADET L., BRAATZ F., GERNER H.J., WOLF S.I., "Kinematics and kinetics with an adaptive ankle foot system during stair ambulation of transtibial amputees", *Gait and Posture*, vol. 30, no. 3, 2009, pp. 356–363.

[AND 03] ANDERSON F.C., PANDY M.G., "Individual muscle contributions to support in normal walking", *Gait and Posture*, vol. 17, no. 2, 2003, pp. 159–169.

[AOY 07] AOYAGI D., ICHINOSE W.E., HARKEMA S.J., REINKENSMEYER D.J., BOBROW J.E., "A robot and control algorithm that can synchronously assist in naturalistic motion during body-weight-supported gait training following neurologic injury", *IEEE Transactions on Neural Systems and Rehabilitation Engineering*, vol. 15, no. 3, 2007, pp. 387–400.

[BAN 09] BANALA S.K., KIM S.H. *et al.*, "Robot assisted gait training with active leg exoskeleton (ALEX)", *IEEE Transactions on Neural Systems and Rehabilitation Engineering*, vol. 17, no. 1, 2009, pp. 2–8.

[BAR 06] BARBEAU H., NADEAU S. *et al.*, "Physical determinants, emerging concepts, and training approaches in gait of individuals with spinal cord injury", *Journal of Neurotrauma*, vol. 23, no. 3–4, 2006, pp. 571–585.

[BAY 05] BAYAT R., BARBEAU H. *et al.*, "Speed and temporal-distance adaptations during treadmill and overground walking following stroke", *Neurorehabilitation and Neural Repair*, vol. 19, no. 2, 2005, pp. 115–124.

[BEH 00] BEHRMAN A.L., HARKEMA S.J., "Locomotor training after human spinal cord injury: A series of case studies", *Physical Therapy*, vol. 80, no. 7, 2000, pp. 688–700.

[BOL 95] BOLMSJO G., NEVERYD H. *et al.*, "Robotics in rehabilitation", *IEEE Transactions on Rehabilitation Engineering*, vol. 3, no. 1, 1995, pp. 77–83.

[CAI 06] CAI L.L., FONG A.J., OTOSHI C.K., LIANG Y., BURDICK J.W., ROY R.R., EDGERTON V.R., "Implications of assist-as-needed robotic step training after a complete spinal cord injury on intrinsic strategies of motor learning", *Journal of NeuroScience*, vol. 20, no. 41, 2006, pp. 10564–10568.

[CAM 09] CAMPOLO D., ACCOTO D. *et al.*, "Intrinsic constraints of neural origin: Assessment and application to rehabilitation robotics", *IEEE Transactions on Robotics*, vol. 25, no. 3, 2009, pp. 492–501.

[COL 00] COLOMBO G., JOERG M. *et al.*, "Treadmill training of paraplegic patients using a robotic orthosis", *Journal of Rehabilitation Research and Development*, vol. 37, no. 6, 2000, pp. 693–700.

[COL 05] COLOMBO R., PISANO F. *et al.*, "Robotic techniques for upper limb evaluation and rehabilitation of stroke patients", *IEEE Transactions on Neural Systems and Rehabilitation Engineering*, vol. 13, no. 3, 2005, pp. 311–324.

[DUB 79] DUBOWSKY S., DESFORGES D.T., "Application of model-referenced adaptive control to robotic manipulators", *Journal of Dynamic Systems, Measurement and Control, Transactions of the ASME*, vol. 101, no. 3, 1979, pp. 193–200.

[EMK 05] EMKEN J.L., REINKENSMEYER D.J., "Robot-enhanced motor learning: accelerating internal model formation during locomotion by transient dynamic amplification", *IEEE Transactions on Neural Systems and Rehabilitation Engineering*, vol. 13, no. 1, 2005, pp. 33–39.

[EMK 06] EMKEN J.L., WYNNE J.H. *et al.*, "A robotic device for manipulating human stepping", *IEEE Transactions on Robotics*, vol. 22, no. 1, 2006, pp. 185–189.

[EMK 07] EMKEN J.L., BENITEZ R. *et al.*, "Human-robot cooperative movement training: Larning a novel sensory motor transformation during walking with robotic assistance-as-needed", *Journal of NeuroEngineering and Rehabilitation*, vol. 4, 2007.

[EMK 08] EMKEN J.L., HARKEMA S.J. *et al.*, "Feasibility of manual teach-and-replay and continuous impedance shaping for robotic locomotor training following spinal cord injury", *IEEE Transactions on Biomedical Engineering*, vol. 55, no. 1, 2008, pp. 322–334.

[ERD 10] ERDEN M.S., TOMIYAMA T., "Human-intent detection and physically interactive control of a robot without force sensors", *IEEE Transactions on Robotics*, vol. 26, no. 2, 2010, pp. 370–382.

[FER 05] FERRIS D.P., CZERNIECKI J.M. *et al.*, "An ankle-foot orthosis powered by artificial pneumatic muscles", *Journal of Applied Biomechanics*, vol. 21, no. 2, 2005, pp. 189–197.

[FIN 91] FINCH L., BARBEAU H. *et al.*, "Influence of body weight support on normal human gait: Development of a gait retraining strategy", *Physical Therapy*, vol. 71, no. 11, 1991, pp. 842–856.

[HAS 97] HASSID E., ROSE D. *et al.*, "Improved gait symmetry in hemiparetic stroke patients induced during body weight-supported treadmill stepping", *Journal of Neurologic Rehabilitation*, vol. 11, no. 1, 1997, pp. 21–26.

[HES 95] HESSE S., BERTELT C. *et al.*, "Treadmill training with partial body weight support compared with physiotherapy in nonambulatory hemiparetic patients", *Stroke,* vol. 26, no. 6, 1995, pp. 976–981.

[HES 99] HESSE S., KONRAD M. *et al.*, "Treadmill walking with partial body weight support versus floor walking in hemiparetic subjects", *Archives of Physical Medicine and Rehabilitation*, vol. 80, no. 4, 1999, pp. 421–427.

[HOG 85] HOGAN N., "Impedance control: An approach to manipulation", *Journal of Dynamic Systems, Measurement, and Control*, vol. 107, 1985, pp. 1–23.

[HOG 00] HOGAN N., "Guest editorial: Rehabilitation applications of robotic technology", *Journal of Rehabilitation Research and Development*, vol. 37, no. 6, 2000, pp. 9–10.

[HOG 06] HOGAN N., KREBS H.I. *et al.*, "Motions or muscles? Some behavioral factors underlying robotic assistance of motor recovery", *Journal of Rehabilitation Research and Development*, vol. 43, no. 5, 2006, pp. 605–618.

[HOO 02] HOOGEN J., RIENER R. *et al.*, "Control aspects of a robotic haptic interface for kinesthetic knee joint simulation", *Control Engineering Practice*, vol. 10, no. 11, 2002, pp. 1301–1308.

[JEZ 04] JEZERNIK S., COLOMBO G. *et al.*, "Automatic gait-pattern adaptation algorithms for rehabilitation with a 4-DOF robotic orthosis", *IEEE Transactions on Robotics and Automation*, vol. 20, no. 3, 2004, pp. 574–582.

[KAE 05] KAELIN-LANE A., SAWAKI L. *et al.*, "Role of voluntary drive in encoding an elementary motor memory", *Journal of Neurophysiology*, vol. 93, no. 2, 2005, pp. 1099–1103.

[KAH 06] KAHN L.E., LUM P.S. *et al.*, "Robot-assisted movement training for the stroke-impaired arm: Does it matter what the robot does?" *Journal of Rehabilitation Research and Development*, vol. 43, no. 5, 2006, pp. 619–629.

[KAO 10] KAO P.C., LEWIS C.L. *et al.*, "Joint kinetic response during unexpectedly reduced plantar flexor torque provided by a robotic ankle exoskeleton during walking", *Journal of Biomechanics*, vol. 43, no. 7, 2010, pp. 1401–1407.

[KWA 97] KWAKKEL G., WAGENAAR R.C. *et al.*, "Effects of intensity of rehabilitation after stroke: A research synthesis", *Stroke*, vol. 28, no. 8, 1997, pp. 1550–1556.

[LAU 01] LAUFER Y., DICKSTEIN R. *et al.*, "The effect of treadmill training on the ambulation of stroke survivors in the early stages of rehabilitation: A randomized study", *Journal of Rehabilitation Research and Development*, vol. 38, no. 1, 2001, pp. 69–78.

[LUM 93] LUM P.S., REINKENSMEYER D.J. *et al.*, "Robotic assist devices for bimanual physical therapy: preliminary experiments", *IEEE Transactions on Rehabilitation Engineering*, vol. 3, no. 2, 1993, pp. 185–191.

[LUM 95] LUM P.S., LEHMAN S.L. *et al.*, "Bimanual lifting rehabilitator: an adaptive machine for therapy of stroke patients", *IEEE Transactions on Rehabilitation Engineering*, vol. 3, no. 2, 1995, pp. 166–173.

[LUM 02] LUM P.S., BURGAR C.G. *et al.*, "Robot-assisted movement training compared with conventional therapy techniques for the rehabilitation of upper-limb motor function after stroke", *Archives of Physical Medicine and Rehabilitation*, vol. 83, no. 7, 2002, pp. 952–959.

[LUM 06] LUM P.S., BURGAR C.G. *et al.*, "MIME robotic device for upper-limb neurorehabilitation in subacute stroke subjects: A follow-up study", *Journal of Rehabilitation Research and Development*, vol. 43, no. 5, 2006, pp. 631–642.

[MUL 10] MULROY S.J., KLASSEN T. *et al.*, "Gait parameters associated with responsiveness to treadmill training with body-weight support after stroke: an exploratory study", *Physical Therapy*, vol. 90, no. 2, 2010, pp. 209–223.

[MUR 85] MURRAY M.P., SPURR G.B. *et al.*, "Treadmill vs. floor walking: kinematics, electromyogram, and heart rate", *Journal of Applied Physiology*, vol. 59, no. 1, 1985, pp. 87–91.

[NAP 89] NAPPER S.A., SEAMAN R.L., "Applications of robots in rehabilitation", *Robotics and Autonomous Systems*, vol. 5, no. 3, 1989, pp. 227–239.

[PAT 06] PATTON J.L., STOYKOV M.E. *et al.*, "Evaluation of robotic training forces that either enhance or reduce error in chronic hemiparetic stroke survivors", *Experimental Brain Research*, vol. 168, no. 3, 2006, pp. 368–383.

[PAT 07] PATTERSON S.L., FORRESTER L.W. *et al.*, "Determinants of walking function after stroke: differences by deficit severity", *Archives of Physical Medicine and Rehabilitation*, vol. 88, no. 1, 2007, pp. 115–119.

[PAT 08] PATTERSON S.L., RODGERS M.M. *et al.*, "Effect of treadmill exercise training on spatial and temporal gait parameters in subjects with chronic stroke: A preliminary report", *Journal of Rehabilitation Research and Development*, vol. 45, no. 2, 2008, pp. 221–228.

[REY 03] REYNOLDS D.B., REPPERGER D.W., PHILLIPS C.A., BANDRY G., "Modeling the dynamic characteristics of pneumatic muscle", *Annals of Biomedical Engineering*, vol. 31, 2003, pp. 310–317.

[RIE 05] RIENER R., LUNENBURGER L. *et al.*, "Patient-cooperative strategies for robot-aided treadmill training: first experimental results", *IEEE Transactions on Neural Systems and Rehabilitation Engineering*, vol. 13, no. 3, 2005, pp. 380–394.

[ROY 09] ROY A., KREBS H.I. *et al.*, "Robot-aided neurorehabilitation: a novel robot for ankle rehabilitation", *IEEE Transactions on Robotics*, vol. 25, no. 3, 2009, pp. 569–582.

[SEI 81] SEIREG A., GRUNDMANN J.G., "Design of a multitask exoskeletal walking device for paraplegics", in *Biomechanics of Medical Devices*, Marcel Dekker, Inc., New York, 1981, pp. 569–644.

[SET 07] SETH A., PANDY M.G., "A neuromusculoskeletal tracking method for estimating individual muscle forces in human movement", *Journal of Biomechanics*, vol. 40, no. 2, 2007, pp. 356–366.

[SHE 06] SHELBURNE K.B., TORRY M.R. *et al.*, "Contributions of muscles, ligaments, and the ground-reaction force to tibiofemoral joint loading during normal gait", *Journal of Orthopaedic Research*, vol. 24, no. 10, 2006, pp. 1983–1990.

[SLO 91] SLOTINE J.-J.E., LI W., *Applied Nonlinear Control*, Prentice-Hall, Englewood Cliffs, NJ, USA, 1991.

[SUG 07] SUGAR T.G., HE J. *et al.*, "Design and control of RUPERT: A device for robotic upper extremity repetitive therapy", *IEEE Transactions on Neural Systems and Rehabilitation Engineering*, vol. 15, no. 3, 2007, pp. 336–346.

[TEI 01] TEIXEIRA DA CUNHA FILHO I., LIM P.A.C. *et al.*, "A comparison of regular rehabilitation and regular rehabilitation with supported treadmill ambulation training for acute stroke patients", *Journal of Rehabilitation Research and Development*, vol. 38, no. 2, 2001, pp. 245–255.

[VAL 09] VALLERY H., VAN ASSELDONK E.H.F. *et al.*, "Reference trajectory generation for rehabilitation robots: Complementary limb motion estimation", *IEEE Transactions on Neural Systems and Rehabilitation Engineering*, vol. 17, no. 1, 2009, pp. 23–30.

[VEN 06] VENEMAN J.F., EKKELENKAMP R. *et al.*, "A series elastic- and bowden-cable-based actuation system for use as torque actuator in exoskeleton-type robots", *International Journal of Robotics Research*, vol. 25, no. 3, 2006, pp. 261–281.

[VEN 07] VENEMAN J.F., KRUIDHOF R. *et al.* "Design and evaluation of the LOPES exoskeleton robot for interactive gait rehabilitation", *IEEE Transactions on Neural Systems and Rehabilitation Engineering*, vol. 15, no. 3, 2007, pp. 379–386.

[VIS 98] VISINTIN M., BARBEAU H. *et al.*, "A new approach to retrain gait in stroke patients through body weight support and treadmill stimulation", *Stroke*, vol. 29, no. 6, 1998, pp. 1122–1128.

[VUK 74] VUKOBRATOVIC M., HRISTIC D. *et al.*, "Development of active anthropomorphic exoskeletons", *Medical and Biological Engineering*, vol. 12, no. 1, 1974, pp. 66–80.

[WIN 90] WINTER D.A., *Biomechanics and Motor Control of Human Movement*, John Wiley & Sons Inc., New York, 1990.

[WIN 91] WINTER D.A., *The Biomechanics and Motor Control of Human Gait: Normal, Elderly and Pathological*, University of Waterloo Press, Waterloo, 1991.

[WOL 08] WOLBRECHT E.T., CHAN V. *et al.*, "Optimizing compliant, model-based robotic assistance to promote neurorehabilitation", *IEEE Transactions on Neural Systems and Rehabilitation Engineering*, vol. 16, no. 3, 2008, pp. 286–297.

[ZIS 07] ZISSIMOPOULOS A., FATONE S. *et al.*, "Biomechanical and energetic effects of a stance-control orthotic knee joint", *Journal of Rehabilitation Research and Development*, vol. 44, no. 4, 2007, pp. 503–513.

Chapter 7

Intelligent Assistive Knee Exoskeleton

This chapter explores the modeling, control, and implementation of pneumatic artificial muscles as actuators for a lower limb exoskeleton. The exoskeleton is aimed to serve as an assistive device to aid disabled persons. The intelligence of the exoskeleton is derived from the user's own myoelectric signals. These signals are processed and used as the reference for the motion controller.

7.1. Introduction

7.1.1. *Background on assistive devices*

The percentage of elderly people in the present world is increasing at an alarming rate. As Table 7.1 shows, in many developed countries the percentage of people over 65 years is close to 20% [STA 10]. As the population of a country ages, there is a corresponding increase in the number of people with physical disabilities. As an example, Japan, the world's third largest economy, is experiencing a phenomenon that is the first of its kind in the world. In 2005 the elderly made up 20% of its total population and it is predicted that by 2030 this percentage will grow to about 30% [OHN 06]. The immense strain that this situation will place on the economy can only be alleviated if active participation of the elderly in society could be prolonged.

Another leading cause of adult disability is cerebrovascular accidents, otherwise known as a stroke. In the United States it is the leading cause of permanent disability. According to the 2010 report from the American Heart Association, on

Chapter written by Mervin CHANDRAPAL, Xiaoqi CHEN and Wenhui WANG.

average every 40 seconds a person suffers from a stroke [LLO 10]. An estimated 795,000 people suffer from a stroke each year and those who survive require extensive and expensive rehabilitation treatment to regain sufficient motor control for independent living [HOO 09].

Country	(%) 2005			(%) 2030 (projection)		
	0–14 years	15–64	65 >	0–14 years	15–64	65 >
Japan	13.7	65.8	20.1	9.7	58.5	31.8
Germany	14.3	66.8	18.9	12.5	59.3	28.2
Italy	14.2	66.2	19.6	12.3	60.9	26.8
France	18.4	65.1	16.5	16.4	59.3	24.3
Korea, Rep	19.1	71.6	9.3	12.6	64.2	23.2
Canada	17.6	69.3	13.1	16.1	61.2	22.7
Sweden	17.4	65.4	17.2	17.0	60.3	22.6
U.K	18.0	65.9	16.1	17.2	62.0	20.9
U.S.A	20.8	66.8	12.4	18.0	62.3	19.8
Russia	15.1	71.1	13.8	15.2	65.4	19.4
China	22.0	70.4	7.6	16.9	67.2	15.9
Brazil	27.5	66.3	6.2	17.0	69.3	13.7
India	33.1	62.3	4.6	22.8	68.8	8.4

Table 7.1. *Age structure of population by country. (Source: Statistics Bureau, MIC; Ministry of Health, Labour and Welfare; United Nations [STA 10])*

With the advent of robotic and mechatronic technologies, tools to empower the disabled can be developed to bridge the treatment gap and provide better, cheaper, and more efficient solutions. Therein lies the goal of assistive and rehabilitation engineering, to improve the quality of life of people with disabilities and the elderly, and to provide sustained rehabilitative therapy. Assistive devices can provide a disabled person with an unprecedented degree of independence to perform activities of daily living such as walking, climbing stairs, and sit to stand movements.

Rehabilitation devices, however, are designed to mimic the movements of a professional therapist and are able to ensure repeatability and increase the frequency of therapy sessions [SEN 09]. When intelligence is incorporated into a rehabilitation robot the progress of the patient can be monitored and the training tuned to suit the needs of the patient.

With such a vast possibility for robotic application in the field of assistance and rehabilitation, it is no wonder that in the last 10 years there has been an explosive growth in research and application of assistive and rehabilitation robotics [KRE 06].

Assistive/rehabilitative (AR) devices can broadly be classified into two main categories, upper extremity and lower extremity. These can be further classified into active and passive devices. Passive AR robots are capable of only resisting forces exerted by the user, such as the walking cane. On the other hand, active devices incorporate actuators to supply and supplement the lack of force from the user. Usually some form of control architecture is integrated to ensure the performance and safety of the device. The following chapter in particular will discuss lower limb assistive devices.

7.1.2. *Lower extremity AR devices*

The evolution of the lower limb exoskeleton can be traced as far back as the 1970s [VUK 90]. The motivation for such devices range from restoration of locomotion for the disabled and elderly to force augmentation for military and industrial applications. In the early days, computational power and actuator size significantly hampered the practicality of a powered lower limb exoskeleton; however, with the advent of smaller and lighter actuators coupled with the exponential increase in computational capability, a practically powered exoskeleton is almost within grasp.

The following overview of the current state of art would aptly demonstrate the tremendous rise in powered exoskeleton research.

7.1.2.1. *AKROD*

Research at Northeastern University in Boston has resulted in an innovative approach to lower limb rehabilitation [WEI 07]. The device named AKROD (active knee rehabilitation orthotic device) aims to enhance gait retraining and improve orthotic intervention in the home and community settings (Figure 7.1). As an answer to the drastically limited contact time during gait retraining, the fabricated device could be worn by patients through daily activities. The constant reinforcement prevents compensatory gait and increases the effectiveness of gait retraining.

The device corrects knee hyperextension and stiff-legged gait pattern by means of a resistive variable damper. Mounted on the knee joint of the orthosis is an electro-rheological fluid (ERF) brake [NIK 06]. The fluid serves as a viscous damper to control the resistive torque of the brake. The viscosity of the fluid is increased in the presence of an electric field. The response time of the fluid to change consistency from a liquid to a viscoelastic gel is in the order of milliseconds.

Figure 7.1. *AKROD testing by an able bodied user [WEI 07] (© 2007 IEEE)*

The closed loop control of the system was achieved using an adaptive proportional-integral torque controller.

Initial human testing with the orthosis shows encouraging results. The controller designed is able to accurately regulate the resistive torque and velocity whilst maintaining a sufficient degree of comfort to the user.

7.1.2.2. Berkley lower extremity exoskeleton (BLEEX)

The BLEEX project has been an ongoing research at the University of California for a number of years and two versions currently exist. The motivation behind the project is to augment human strength. The device could potentially be used by fire fighters, disaster relief workers, and soldiers to carry heavier loads, further than otherwise possible (Figure 7.2).

The device is powered by hydraulic actuators at the hip (flexion and extension), knee (flexion) and ankle (plantarflexion and dorsiflexion[1]) along the sagittal plane. Sensors mounted on the soles provide force feedback [HUA 05, KAZ 05].

However as BLEEX was designed to be used by an able-bodied person, the control architecture is structured so as to minimize impedance to the human user. There are no algorithms implemented to control postural stability, this has to be managed solely by the user. The sole purpose of the system is to bear the additional

---

1 Plantarflexion is when the forefoot is moved away from the body and the opposing movement when the forefoot is pulled toward the body is known as dorsiflexion.

load on the user. Autonomous operation is achieved through a small fuel engine that powers the hydraulic components and the onboard computer.

Figure 7.2. *BLEEX (with permission from H. Kazerooni, UC, Berkeley)*

7.1.2.3. *Pneumatically powered knee-ankle-foot orthosis (KAFO) – University of Michigan*

In an effort to understand the effects of a powered exoskeleton on normal human gait, pneumatically powered knee-ankle-foot (KAFO) [SAW 09] and ankle-foot orthosis (AFO) [FER 05, SAW 05] were developed at the University of Michigan. The research which spans for almost a decade now is centered on the application of torque (through pneumatic artificial muscles (PAM)) across the ankle and knee joints (Figure 7.3).

The orthoses were constructed from carbon fiber shells to custom fit the user's leg [GOR 06, FER 06]. Surface electromyography (sEMG) signals from the tibialis anterior[2] and soleus[3] were used as control signals for ankle dorsiflexion and plantar flexion in the AFO. In addition to the sEMG signals from muscles at the ankle,

---

2 The tibialis anterior muscle is the most medial muscle at the front of the leg. It is responsible for dorsiflexing and inverting the foot.
3 The soleus is a powerful muscle in the back part of the calf. It runs from just below the knee to the heel. The action of the soleus is the plantarflexion of the foot.

sEMG from the vastus lateralis and medial hamstrings were used for knee extension and flexion in the KAFO. The PAM force delivered was proportional to the rectified and filtered sEMG signals (i.e. proportional myoelectric control).

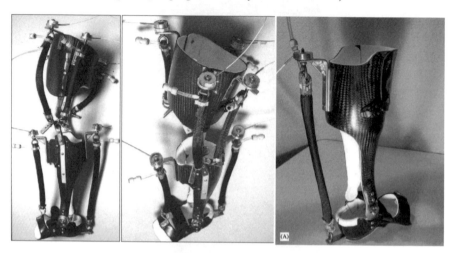

Figure 7.3. *Knee-ankle-foot orthosis (right) and ankle-foot orthosis (left) [SAW 09]. Image used with permission from BioMed Central (right image) and ELSEVIER B.V (left image)*

The latest publication on the research concluded that while the powered exoskeleton (AFO) delivered about 63% of the average ankle joint mechanical power during walking, this did not result in a proportional reduction of net metabolic power (about 10%). The authors suggest that metabolic savings when using a powered exoskeleton are less in joints with considerable elastic compliance (as in the ankle joint). Larger metabolic savings may be possible in joints that rely on power production due to positive muscle work, such as in the knee joint [SAW 08].

7.1.2.4. *Hybrid assistive leg (HAL)*

Another long standing research carried out by the University of Tsukuba (Japan) in cooperation with Cyberdyne Systems Company, is the hybrid assistive leg (HAL) project. The HAL suit is a wearable robot designed for human strength augmentation and to increase the quality of life for the elderly (Figure 7.4). More recently the work has been extended to provide walking motion support for persons with hemiplegia and gait rehabilitation for persons with incomplete spinal cord injury [KAW 09].

The exoskeleton is powered by dc servo drive motors at the hip and knee joint and in HAL 5 the elbow and shoulder joints are also powered. Various control systems have been developed to guide the movements of the exoskeleton with respect to the human user. The first exoskeletons utilized EMG feedback from the extensor muscles to determine the operator's intentions [KAW 03b]. Later phase sequence algorithms were utilized to move the exoskeleton through a predefined motion based on the intention of the operator [KAW 04, KAW 03a]. A hybrid controller incorporating both the biological signals from the operator and motion information proved to be the most effective means of matching the viscoelasticity of the HAL suit to that of skeletal muscle [HAY 05].

To date, the HAL exoskeleton is recognized as one of the very few mobile, powered exoskeletons that are commercially available. However, research is still being carried out to diversify the application of the HAL suit [KAW 09].

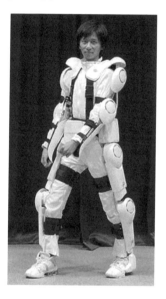

Figure 7.4. *A student demonstrating the HAL-5 powered exoskeleton [UBE 07]*

This review of powered exoskeleton research is obviously not an exhaustive list of the research that is currently being carried out; however, those highlighted here have an immediate relation to the work that will be discussed in this chapter. Much inspiration has been drawn from the HAL project and the research at the University of Michigan.

7.1.2.5. *Actuators for lower limb exoskeleton*

There are various constraints imposed on a powered lower limb assistive device. The weight, type of actuator, power supply, range of motion, safety, low impedance and duration of use all play crucial roles. Common types of actuators utilized are electric (DC), hydraulic, and pneumatic drives. Each of these possesses its own advantages and disadvantages.

The electric drives (servo motors) have the advantage of being low cost, relatively easy to attain position and velocity control, and produce very low noise during operation [LOW 06]. On the other hand, the power and torque-to-weight ratio of a servo motor is significantly lower when compared with other types of actuation [CAL 95]. The main drawback of electric drives is the lack of compliance in the actuator. This last characteristic is of considerable importance as it affects the safety of the user. In [PRA 04] an electric servo motor is used in the construction of a series elastic actuator. Low impedance is achieved in this device through complex gearing.

Hydraulic actuators are generally more suitable for an industrial setting due to the high power-to-weight ratio, strength, and rigidity. The last two characteristics though ideal on the shop floor, are not suitable in an AR exoskeleton. Furthermore, hydraulic actuators suffer from high maintenance cost and the possibility of oil leaks.

Conversely, pneumatic actuators whilst possessing a similar high power-to-weight ratio, do not come with the maintenance and leakage issue that exist in hydraulic actuators. Pneumatic actuators are low in cost, have fast response time, and are inherently compliant and backdrivable, thus proving to be a good choice where there is human interaction [DAE 02].

Pneumatic artificial muscles (PAM) manage to provide an excellent "hybrid" that inherits the compliancy and backdrivability of a linear pneumatic actuator but not the weight and bulk. PAM's are therefore ideal for powered AR exoskeletons.

PAM actuators have already been utilized since the 1950s in many AR devices [TON 00]. A significant portion of the research however concerns robotic locomotion, while these are not directly related to this work they do demonstrate the potential of PAM actuators [LIL 03, PAC 97, TAK 06]. The only significant drawback of the PAM actuator is the nonlinearity inherent in the system due to its construction and the compressibility of air. Classical control methods lack the ability to accurately manipulate the actuator. The modeling and control of the PAM actuator therefore is our focus and will be discussed in greater depth in the following sections.

7.2. Overview of knee exoskeleton system

The contribution of this work is the design and development of a pneumatic assistive knee exoskeleton. The basic principle behind the operation of the exoskeleton is the augmentation of the skeletal muscle force produced by the user. The fundamental control architecture of the exoskeleton is depicted in Figure 7.5.

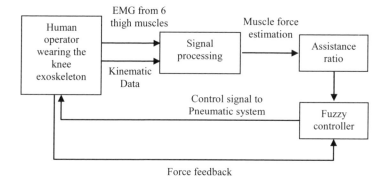

Figure 7.5. *Exoskeleton control architecture*

Surface myoelectric signals from the knee extensor and flexor muscles provide the control link between the user and the exoskeleton. The sEMG signals are obtained through six bipolar electrodes attached to front and rear of the thigh. A preamplifier circuit mounted on the electrode conditions the signals before it is fed into a mapping algorithm. The purpose of the algorithm is to determine the degree of force exerted by the skeletal muscle and calculate the appropriate PAM force depending on the set *assistance ratio*. The encoder mounted on the exoskeleton provides kinematic feedback of the knee joint.

The force difference between the muscle and the PAM is provided as the reference signal to the motion controller. A multiple-input single-output self-organizing fuzzy controller is utilized as the motion controller. The inputs to the controller are the error (e) and the change in error ($\dot{e}$), and the output is the duty cycle of the PWM pulse.

The output from the motion controller regulates the valve opening of a high-speed (on/off) pneumatic solenoid valve. This is accomplished by varying the duty cycle of the 100Hz PWM pulse. Through this method the mass flow rate of air flowing into the PAM can be accurately controlled. Two valves are used to control the flow of air into the PAM, one for inflation and the other for deflation.

It is important to mention that the current exoskeleton design does not include any algorithms to ensure postural stability or to determine the user's complete intended motion (e.g. sit-to-stand or ascending a flight of stairs). The PAMs are intended to operate in parallel to their organic counterparts and as such they augment rather than direct the movement. Since the basic control signal is derived from the sEMG signals, the high level controller in this exoskeleton system is the human motor system (cerebral cortex-brain stem-spinal cord), whereas the low level controller is the PAM motion controller.

Research conducted at the University of Michigan (section 7.1.2.3) has demonstrated that the human motor system is capable of changing muscle activation pattern in response to external support provided during walking. The objective therefore is to allow the motor system to control the coordination of the intended trajectory while the exoskeleton provides the assistance needed to do so. In this sense the exoskeleton is a seamless extension to the human motor system.

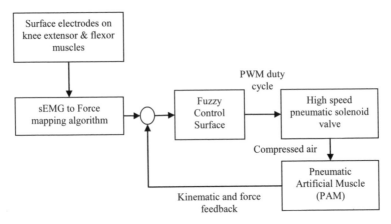

Figure 7.6. *Exoskeleton control block diagram*

The control block diagram of the exoskeleton is illustrated in Figure 7.6, for the first and second generation knee exoskeleton prototypes, shown in Figure 7.7. The following sections in this chapter will elaborate on the various components of the exoskeleton as shown in the block diagram.

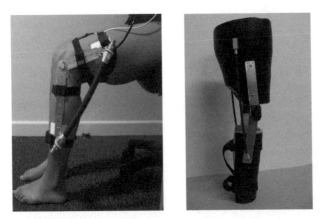

Figure 7.7. *First (right) and second (left) generation exoskeleton prototypes. (One 10 mm diameter PAM was used in the first prototype and two 5 mm were used in the second)*

7.3. Modeling and control of pneumatic artificial muscle (PAM)

7.3.1. *Background*

The pneumatic artificial muscle otherwise known as the Mckibben muscle is a variant of the traditional pneumatic actuator. The invention of the muscle is attributed to Joseph L. Mckibben who in the 1950s utilized it to actuate an arm orthosis [SCH 61]. After initial popularity the actuator fell out of favor as electric drives rose in popularity in the 1960s. In the 1980s, the Japanese tire manufacturer Bridgestone reinvented a more powerful version of the PAM for a painting application called rubbertuator [INO 88]. However in the past decade there has been a renewed interest in the PAM actuator [ANH 08, BAL 03, BON 09, CHO 06, MIN 97, PAC 97, SIT 08, XIA 08] as a result of its inherent compliance and its properties that mimic skeletal muscle.

The Mckibben muscle falls under the category of braided muscles which essentially are an inflatable membrane enclosed within a braid. The sleeve or braid is a helical mesh that runs the length of the PAM. The PAM operates at an overpressure, that is the PAM exerts force on a load when supplied with compressed air (usually about 400–600 kPa). When pressurized the inner gas-tight elastic tube expands radially exerting force on the braid. The tension in the braid fiber balances the force exerted and this tension is summed at the endpoints of the PAM. The force then can be transferred to an external load. A CAD model of the internal construction of the PAM is given in Figure 7.8.

Figure 7.8. *CAD model of the Festo fluidic muscle [FES 10]. The internal braid (and braid angle) is clearly visible encasing the inflatable membrane*

Early researches have opted to construct their own muscles based on the general design of the PAM; the problem that soon arises from this is the inability to reliably reproduce the characteristics of each PAM. To avoid this issue we have opted to utilize a commercially available muscle. One manufacturer of commercial PAM is the German company FESTO. They produce Mckibben muscles of various diameters (5 mm, 10 mm, 20 mm, and 40 mm) with known force and pressure characteristics (Figure 7.9). This type of PAM is capable of a maximum contraction of up to 25% of its original length. The detail characteristic data provided by the manufacturer provides a good foundation for further work.

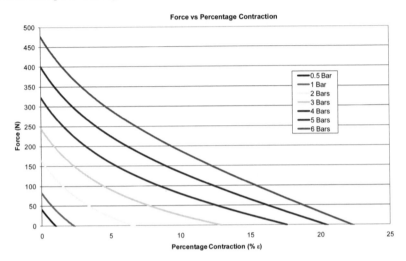

Figure 7.9. *Force vs. percentage contraction for different pressures within the PAM (graph is for 10mm diameter PAM)*

7.3.2. *Characteristic of the PAM*

The ability of the PAM to inflate during contraction whilst maintaining a general cylindrical shape is the result of bias angle in the helical braid. The bias angle or braid angle is the angle between the PAM axis and the braid thread. By employing a weak enough initial angle the braid allows the expansion of the inner tube under pressure thereby converting circumferential pressure to axial force [TON 00].

Daerden and Lefeber [DAE 02] conducted two simple experiments to examine the operation of the PAM in which a PAM was fixed vertically at one end and had a mass hanging from the other end. In the first experiment the hanging mass was kept constant and the pressure within the PAM was gradually increased. In the second experiment the pressure within the PAM was kept constant and the mass was gradually reduced. The experiments led Daerden to propose rules to describe the behavior of the PAM:

1. The contraction of the PAM is realized by increasing its volume.

2. Contraction against a constant load can be achieved by increasing the pneumatic pressure within the PAM.

3. As the loading on a PAM is decreased, its contraction will occur at a constant pressure.

4. The contraction of the PAM has an upper limit at which point it reaches its maximum enclosed volume and exerts zero pulling force.

This last characteristic is of particular interest because in traditional linear pneumatic drives the force exerted is only proportional to the piston cross-sectional area and the pressure within the piston chamber. However in a PAM the force exerted is proportional to both the pressure and the muscle contraction.

PAMs possess many characteristics that are desirable in an exoskeleton actuator. The power-to-weight ratio of most PAMs is in the order of kN/kg. The PAM that is utilized in this work (MAS-10-N290) has a force-to-weight ratio of approximately 3.5 kN/kg [FES 03]. PAMs also exhibit inherent compliance and naturally damped dynamic response. The compliance is the direct result of the compressibility of air and the elasticity of the inflatable membrane while the damped response is due to the nonlinear kinetic friction intrinsic in the outer braid [DAE 02]. This compliant behavior is highly useful where there is close human–machine interaction.

Additionally, PAMs also bear a close resemblance to human skeletal muscle. Both are contractile actuators only capable of attractive forces and require an antagonistic setup for bidirectional motion. The decreasing force-contraction characteristic seen in PAMs is similar to the force-contraction characteristic in skeletal muscle. These attributes together with the intrinsic compliance make PAM an ideal extension to the human skeletal muscle.

7.3.3. Models from literature

The compressibility of air and the hysteresis in the PAM actuator result in a highly nonlinear system that is difficult to both model and control. Researchers [DAE 02, MIN 10, TON 00] have attributed the hysteresis losses to thread-on-thread dry friction acting within the PAM braided shell, the friction between the braid and the inner bladder, and the hysteresis of the inner bladder itself.

As a result many models have been developed to approximate the properties of the PAM. Two common models used in literature will be discussed. The first is based on the principle of virtual work and the geometry of the PAM braid [CHO 96, TON 00].

Assuming that the PAM maintains its cylindrical shape during contraction, the principle of virtual work can be applied to determine the force exerted. As the PAM contracts its radius increases from r_0 to r and its length decreases from l_0 to l.

Assuming positive motion in the direction of muscle extension, the virtual work of the pressure force can be expressed by:

$$P\delta V = F\delta l \qquad\qquad [7.1]$$

where V is the change in volume as the result of the increase in radius. Taking into account the current braid angle a, the expression of F, the force exerted by the PAM as a function of the control pressure P and the contraction ratio ε (l_0/l) is shown to be:

$$F(\varepsilon,P) = (\pi r_0^2)P[a(1-\varepsilon)^2 - b], 0 \le \varepsilon \le \varepsilon_{max}$$
$$\varepsilon = (l_0 - l)/l_0, \quad a = 3/\tan^2(a_0), \quad b = 1/\sin^2(a_0) \qquad\qquad [7.2]$$

The model shows the change in force exerted by the PAM; at zero contraction the force is maximum and at maximum contraction the force falls to zero.

The model derived from virtual work requires prior knowledge of the braid angle; it is however impossible to determine the braid angle in a commercial muscle such as the one that has been utilized in the work. It is therefore necessary to take another approach to model the PAM. Situm and Herceg [SIT 08] have modeled a similar PAM from the same manufacturer by approximating the response with a first order lag term. The transfer function is defined as the ratio of the pressure in the actuator to the control signal of the proportional pneumatic valve.

$$\frac{p(s)}{u(s)} = \frac{K_m}{T_m s + 1} \qquad\qquad [7.3]$$

The transfer gain K_m and the time constant T_m are determined experimentally. The pressure is then related to the force exerted based on a physical static model without weave geometries. This equation expresses the force developed by the PAM as a function of only the pressure and contraction ratio.

$$F = K_p P[l_{max}(1-\varepsilon) - l_{min}] \qquad\qquad [7.4]$$

where l_{max} and l_{min} are the relaxed and fully contracted muscle length. The constant K_p which is dependent on the working pressure P is determined experimentally.

The static models described do not take into account the hysteresis inherent in the PAM. Often the hysteresis in the PAM is not explicitly modeled but grouped together as part of the nonlinearity in the actuator. In a recent work by Minh et al.

[MIN 10] the hysteresis in the PAM is explicitly modeled using a Maxwell slip model and used in a feed forward compensator in the control loop.

7.3.4. *Model used*

In this work the PAM is modeled as a quasi-static system with the hysteresis incorporated into the nonlinear characteristic curve of the actuator. The main motivation to model the PAM was to enable the offline tuning of the self-organizing fuzzy controller. It is important to note here that the modeling of the pneumatic subsystems is not essential for the fuzzy controller. The controller is fully capable of tuning itself online as the system operates. However, as a preliminary step the offline (simulation) tuning of the controller was performed to evaluate the performance of the controller. Thus it was necessary to obtain models of the subsystem (PAM-high speed valve-air mass flow rate) to perform simulation. The utilization of the models for the offline tuning of the fuzzy controller will be described in section 7.5.4.

To obtain an empirical approximate model of the PAM, look-up tables instead of a mathematical model is used. It is known from the literature that the force developed by the PAM is proportional to its percentage contraction ($\%\varepsilon$) and the pressure in the PAM. The pressure and percentage contraction in the PAM is in turn proportional to the mass of air in the PAM and the loading on it. Therefore it follows that if the mass of air in the PAM and the loading is known at all times then the $\%\varepsilon$ and the force exerted by the PAM can be determined.

The experimental data points for the look-up table were obtained by attaching a fixed load on the PAM and increasing the pressure within the PAM at set increments. The $\%\varepsilon$ and the volume of air in the muscle (the PAM is approximated to a cylinder and the volume of air contained within is calculated) for each increment in pressure were recorded. The entire procedure is repeated with the loading increased from 0–150 N at 10 N increments. The experiment carried out is similar to those conducted by Daerden and Lefeber that was mentioned earlier.

The reason the PAM data was obtained experimentally rather than by simply using the data provided by the manufacturer is first to acquire the necessary additional information regarding the mass of air in the PAM at different pressures and to also verify the data provided by the manufacturer. The first table (here shown as a 3D surface, Figure 7.10) relates the pressure within the PAM to the mass of air in the PAM and the loading.

By means of these two characteristic curves the $\%\varepsilon$ of the PAM and the pressure within the PAM can be determined if the mass of air entering the PAM is known.

The modeling of the pneumatic valve to determine the mass flow rate will be discussed in section 7.4.

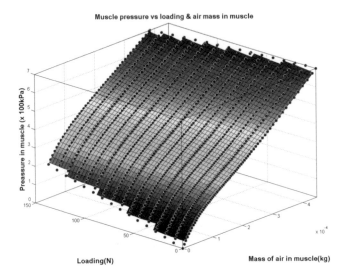

Figure 7.10. *Pressure in PAM (kPa) vs. loading (N) and mass of air in muscle (kg)*

The second surface (Figure 7.11) relates the %ε of the muscle to the pressure within the muscle and the loading. The force can then be calculated as a function of %ε and pressure.

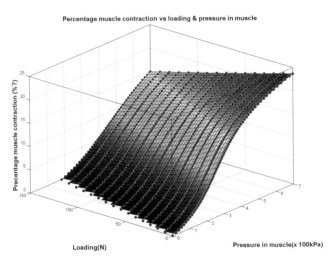

Figure 7.11. *Percentage contraction (%ε) vs. loading (N) and pressure in the muscle (kPa)*

7.4. Modeling of high-speed on/off solenoid valve

In literature the most prevalent way to regulate the pressure within the PAM is through a proportional pressure regulator. The regulator provides a simple and easy means to control the pressure within the PAM by varying the voltage or current. However, pressure regulators are often expensive, bulky, and require constant power supply to function.

Another method proposed in literature is to pulse a high-speed solenoid valve to vary the valve opening [BAR 97, CHE 07, SIT 08, ZHA 08]. The flow equation relating the valve orifice area, A_V to mass flow rate, $\dot{m}$ is:

$$\dot{m} = \begin{cases} C_f A_v C_1 \dfrac{P_u}{\sqrt{T}} & \text{if } \dfrac{P_d}{P_u} \leq P_{cr} \\[2ex] C_f A_v C_2 \dfrac{P_u}{\sqrt{T}} (\dfrac{P_d}{P_u})^{\kappa-1} \sqrt{1-(\dfrac{P_d}{P_u})^{\frac{\kappa-1}{\kappa}}} & \text{if } \dfrac{P_d}{P_u} \geq P_{cr} \end{cases}$$

[7.5]

In this experiment the valve orifice is assumed to be a circle, which is a close approximation of the shape of the orifice and C_f is a non-dimensional discharge coefficient through the orifice. κ is the ratio of specific heats. The mass flow rate is also dependent on the ratio of the pressure before (P_u, upstream pressure) and after the orifice (P_d, downstream pressure). The orientation of the P_u and P_d change depending on whether the PAM is inflating or deflating. During inflation, P_u is the supply pressure of 600 kPa and P_d is the increasing pressure within the PAM. During deflation the opposite is true, P_u is the decreasing pressure within the PAM and P_d is the constant atmospheric pressure.

The constants C_1, C_2, and the critical pressure P_{cr} are calculated as equation [7.6]. For the PAM system the critical pressure ratio P_{cr} for the system is $P_{cr} = 0.528$.

$$C_1 = \sqrt{\frac{\kappa}{R}\left(\frac{2}{\kappa+1}\right)^{\frac{\kappa+1}{\kappa-1}}} \quad C_2 = \sqrt{\frac{\kappa}{R}\left(\frac{2}{\kappa+1}\right)} \quad P_{cr} = \left(\frac{2}{\kappa+1}\right)^{\frac{\kappa}{\kappa-1}}$$

[7.6]

Two high-speed 3/2 way solenoid valves (FESTO MHE2-MS1H-3/2G) with a switching time of approximately 2 ms are used to adjust the orifice opening, and control the pressure within the PAM. One valve is responsible for inflating the PAM and the other is used to deflate the PAM. Thus the valves are in fact used as 2/2 way valves with the exhaust port plugged.

The maximum operating frequency of this valve is 330 Hz (3 ms pulse period); however since the minimum valve switching time is 4 ms the operating frequency was set to 100 Hz (10 ms pulse period). This frequency was chosen to provide a gradual increase in flow rate with regard to the PWM duty cycle. At high frequencies the operation of the high-speed valve is subject to complex electric and magnetic influences. Electrical delay, magnetic delay, and mechanical delay all combine to retard the response of the valve [KAJ 95].

An intuitive attempt to model the dynamics of the on-off valve is proposed by Chen et al. [CHE 07]. The initial model is here adapted and extended to encompass all the possible states of the valve. The initial model is given in equation [7.7] and graphically depicted in Figure 7.12.

$$
\begin{aligned}
&X_i = & &0 \\
&\text{when} & &t \in [(t-1)T_c, (t-1)T_c + t_1] \\
&X_i = & &\frac{X_m}{t_2}[t - (t-1)T_c - t_1] \\
&\text{when} & &t \in [(t-1)T_c + t_1, (t-1)T_c + t_1 + t_2] \\
&X_i = & &X_m \\
&\text{when} & &t \in [(t-1)T_c + t_1 + t_2, (t-1)T_c + T_p + t_3] \\
&X_i = & &\frac{X_m}{t_4}[t - (t-1)T_c - T_p - t_3 - t_4] \\
&\text{when} & &t \in [(t-1)T_c + T_p + t_3, (t-1)T_c + T_p + t_3 + t_4] \\
&X_i = & &0 \\
&\text{when} & &t \in [(t-1)T_c + T_p + t_3 + t_4, iT_c]
\end{aligned}
\qquad [7.7]
$$

where X_i is the spool displacement, X_m the maximum spool displacement, U PWM pulse magnitude, i the number of PWM pulses starting from 1, 2, 3...n, T_c the PWM period, T_p the PWM on period, t_1 electrical delay and magnetic delay (armature picking up time, approx 1 ms), t_2 mechanical delay (spool responding time, approx 1 ms), t_3 electrical delay and magnetic delay (armature take down time, approx 1 ms), t_4 mechanical delay (spool release time, approx 1 ms). The switching on and off time are given by $t_1 + t_2$ (2 ms) and $t_3 + t_4$ (2 ms).

The equation given however is only applicable when $(t_1 + t_2) \leq T_p \leq (T_c - t_3 - t_4)$. In this work this state is referred to as the third state (out of five). To accommodate other possible states four other switching modes were identified and modeled. These states are dependent on the PWM period, and model the characteristics when the PWM period is either too short so that the spool will not fully extend or too long, preventing the full return of the valve spool.

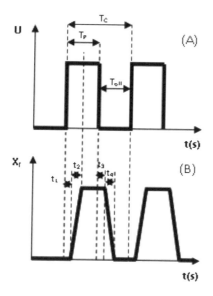

Figure 7.12. *State 3-illustration of spool displacement, PWM pulse (A),*
Spool displacement (B)

7.4.1. *Experimental validation of high-speed valve flow rate*

The flow rate profile of the high-speed valve with PWM control that was obtained through simulation (shown in Figure 7.13) was experimentally verified. Data obtained through simulation illustrates the dead band and the saturation that occurs within the high-speed valve at low and high duty cycles respectively.

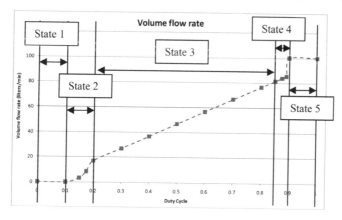

Figure 7.13. *Simulated high-speed valve volume flow rate vs. PWM duty cycle*

The experimental setup to verify the simulation data was carried out using a (MHE2-MS1H-3/2G) valve and a flowmeter (Platon NG). The results obtained show that the model overestimates the flow linearity and maximum flow rate (Figure 7.14). However, the dead band and saturation of the valve are evident from the graph; the discrepancy in the linearity may possibly be due to nonlinear flow profile through the orifice. To better reflect the actual valve flow profile, the discharge coefficient and the maximum valve opening in equation [7.5] were readjusted.

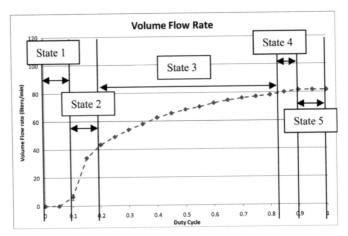

Figure 7.14. *Experimental high-speed valve volume flow rate vs. PWM duty cycle*

7.5. Self-organizing fuzzy control

7.5.1. *Introduction to fuzzy control*

Fuzzy control is the offspring of fuzzy set theory proposed by Lofti A. Zadeh in the 1960s and 1970s [ZAD 65, ZAD 68]. It essentially is a method of intelligent control based on human heuristic knowledge. In fuzzy logic a set is said to have varying degrees of truth ranging from 0 to 1. This is in contrast to the traditional Boolean logic of either 0 or 1. In fact, Boolean logic can be seen as a boundary form of fuzzy numbers and sets.

In everyday life, most things are described with *linguistic variables* and *modifiers*. As an example the linguistic variable height can have linguistic values of short, medium, and tall, and can be modified with, slightly, very, and extremely. These variables do not have a defined boundary, or in other words are "fuzzy". Figure 7.15 gives the fuzzy representation of the set of tall men. The *universe of discourse* (*x* axis) contains all the elements that are considered, in this case the height of men.

Membership

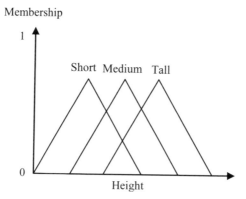

Figure 7.15. *Fuzzy set of men's height*

When applied to control systems, fuzzy logic has the capability to accurately control highly nonlinear systems [JAN 98, KOV 06]. The decision-making process in a fuzzy control system is based around the very common IF-THEN statement.

IF the input is X, THEN the output is Y.

This structure is highly suitable to capture the experience and knowledge of a human operator. These IF-THEN rules are stored in an inference engine and executed based on the input to the system. As a result fuzzy control is an effective tool when dealing with ill-defined systems where the mathematical modeling is poor [JAN 98c, KOV 06]. The fuzzy inference engine operates based on expert human knowledge and experience in place of a detailed model. The complete concept of fuzzy control will be explained with respect to the PAM control system.

7.5.2. *Fuzzy control system for PAM*

First the input and output variables of the fuzzy controller is defined. The fuzzy control system for the PAM is based on a PD-type controller. The input to the controller is the error (e) and the differential of the error (ce). The error is the difference between the reference force and the force generated by then PAM. Figure 7.16 gives the simplified overview of the control system.

The output of the fuzzy controller is the duty cycle (u) of the PWM generator which will in turn regulate the air flow into the PAM. This will enable accurate control of the pressure within the PAM and thereby the force exerted by the PAM. The output duty cycle ranges from −100% (for deflation of the PAM) to 100% (for PAM inflation).

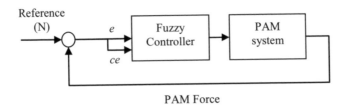

Figure 7.16. *PAM fuzzy control system*

Both input variables and the output variable (Figure 7.17) have seven membership functions in the fuzzy set, over the universe of discourse (−1 to 1). In the case of the output fuzzy set, −1 corresponds to −100% deflation and the opposite, for inflation. The linguistic values [(B)ig, (M)edium, (S)mall, and (Z)ero] modify the (P)ositive and (N)egative input values. Each linguistic value is represented as a *membership function*. A function that ties each element of the *universe* to a membership value is called a *membership function*. The membership functions are biased toward the center to give better control when the error or change-in-error is close to equilibrium.

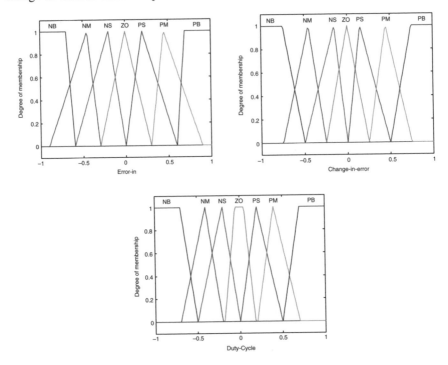

Figure 7.17. *Fuzzy controller output, duty cycle of the PWM generator*

Triangular membership functions are used for the middle membership functions, while trapezoidal ones are used at the two ends. This is the standard and well-established method for an initial fuzzy controller design [JAN 98]. Although the universe is the same for both input variables, the input gains for each variable can be adjusted to give better system response [JAN 98].

The output membership functions have similar shapes to that of the input variables; however, the central function is a trapezoid instead of a triangle to account for the dead band of the high-speed pneumatic valve.

Next, the fuzzy rule base is built based on expert knowledge of the system that is to be controlled. In the case of the PAM, existing rule base from literature is adapted to suit the requirements of the system [CHE 07, MIN 97, ZHA 08]. The rule base used for the fuzzy controller is given in Table 7.2.

Since there are seven membership functions in each input variable, there is one rule for each input combination to ensure that for every possible input combination there is a rule to determine the appropriate action. As an example, referring to Table 7.2 – IF the error is NB (negative big) AND the change in error is PS (positive small) THEN the output duty cycle is NM (negative medium). This rule can be simplified to:

IF e is NB AND ce is PS THEN u is NM

Finally the fuzzy inference engine evaluates the fuzzified inputs based on the rule base and provides a crisp output.

		Error						
		NB	NM	NS	ZO	PS	PM	PB
Change in error	NB	NB	NB	NB	NB	NM	NS	ZO
	NM	NB	NB	NM	NM	NS	ZO	PS
	NS	NB	NM	NM	NS	ZO	PS	PM
	ZO	NB	NM	NS	ZO	PS	PM	PB
	PS	NM	NS	ZO	PS	PM	PM	PB
	PM	NS	ZO	PS	PM	PM	PB	PB
	PB	ZO	PS	PM	PB	PB	PB	PB

Table 7.2. *Fuzzy controller rule base*

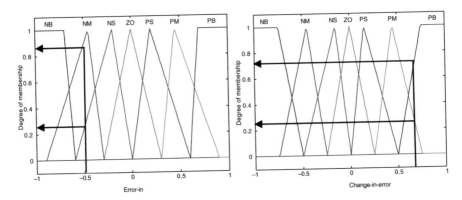

Figure 7.18. *Fuzzification of crisp inputs e=−0.5 and ce=0.6*

The operation of the fuzzy inference engine can then be demonstrated using a simple example. In Figure 7.18 two arbitrary input variables are assumed ($e = -0.5$, $ce = 0.6$). The crisp inputs firstly undergo a fuzzification process. The crisp error input is fuzzified and has about 0.22 degree of membership to the set of NS and 0.82 to the set NM. The crisp change-in-error input is also fuzzified and has degrees of membership of 0.22 to the set PM and 0.66 to the set PB.

This causes all rules with the corresponding membership functions to fire. The combinations of rules that will fire are highlighted in gray in Table 7.2. The firing strength of the rule is determined by the AND method, in this case since the rules are:

$$\text{If } e = u_{e1} \text{ AND } ce = u_{ce1}$$

where u_{e1} and u_{ce1} are the respective activation levels, the *min* operator is used. The operator sets the activation of the antecedent part of the rule equal to the lower activation of the two terms.

The deduction of consequent or the conclusion of the rule (THEN...) is called *fuzzy implication*. The *min* operator is used here also. This operator clips the output (consequent part) membership function based on the activation level of the antecedent part. If more than one rule fires for a given input combination, as in this case, all the clipped consequent parts have to be combined before a crisp output can be determined. This operation is called *fuzzy aggregation*. The *max* operator used for the aggregation will produce an output fuzzy set based on the maximum activation of each output membership function. The final step is the *defuzzification* for the output fuzzy set into a crisp output. The centroid of area method is used for this purpose:

$$u = \frac{\int \mu(x_i) x_i}{\int \mu(x_i)} \qquad [7.8]$$

where $\mu(x_i)$ is the membership value in the membership function and x_i is a running point the continuous universe. The function calculates the output as the weighted average of the elements in the output set [JAN 98, KOV 06]. Figure 7.19 shows the *fuzzification*, *activation*, *implication*, and *defuzzification* of the four rules from the example ($e = -0.5$, $ce = 0.6$). The crisp output from the controller is a duty cycle of 0.2 (20% PAM inflation).

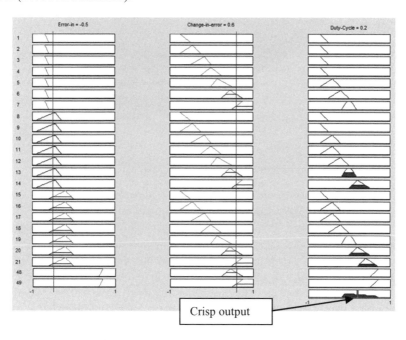

Crisp output

Figure 7.19. *Fuzzy inference engine output based on input of error and change in error (obtained using MATLAB Fuzzy toolbox)*

Once the fuzzy inference engine has been designed it is then possible to determine the crisp output for all possible input combinations. This can be represented in the form of a surface known as the *fuzzy control surface*. The control surface for the PAM is given in Figure 7.20.

However, for control application it is often better to discretize the input *universe* to achieve faster computational speed. Thus the inputs into the fuzzy controller are quantized to increments of ±0.05; this allows the fuzzy control surface to be simplified to a 41 × 41 look-up table. The look-up table then can easily be incorporated into the feed-foward path of the control loop to achieve fast control response.

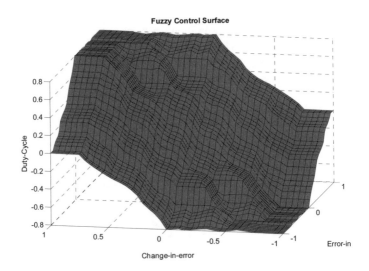

Figure 7.20. *Fuzzy control surface*

7.5.3. *Introduction to self-organizing fuzzy controllers*

The adjustment of the fuzzy membership functions and the rule base is often based on heuristics and expert knowledge. This method though often sufficient does not necessarily always result in a controller that is optimized to affect the desired response for a particular plant. The alternative approach proposed by Mamdani in his pioneering papers [MAM 75, PRO 07] is a fuzzy logic controller that is capable of self-organization iteratively based on the control quality and the desired response of the plant.

One basic structure of a self-organizing controller is a lower level standard-table-based fuzzy controller and a higher level adjustment mechanism (Figure 7.21). The higher lever modifier corrects the values in the table of the lower controller if the current performance does not match the desired.

If the current performance is as desired then the modifier does not have to correct the values in the table; however, if the performance is inadequate then a particular entry in the table has to be adjusted to give the correct control signal. The modifier cannot and should not penalize the current table entry (for current plant output) as there is a delay between the control output and the plant response.

A simplification proposed by Jantzen [JAN 98b] is to correct the control output that occurred in a few samples in the past. The number of samples or the delay-in-penalty "d" corresponds to the time lag of the plant. The modification in the fuzzy table can be expressed as:

$$u_{i-d} = u_{i-d} + \Delta P_i \qquad\qquad [7.9]$$

where i is the current iteration and ΔP_i is the correction factor added to the entry in the table-based controller. From the equation it is evident that this modification algorithm will only succeed if the increase in the control signal corresponds to an increase in the plant output and vice versa. Fortunately this is the case in the PAM system.

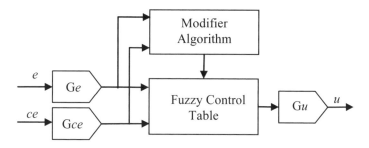

Figure 7.21. *Self-organizing fuzzy controller*

7.5.4. *Practical implementation*

In theory the lower fuzzy table-based controller can initially be zeros and the modifier algorithm should be able to generate the fuzzy table iteratively. Nonetheless, if the fuzzy table is already populated with reasonable values (as described in section 7.5.20) the convergence of the self-organizing fuzzy controllers (SOFC) is considerably faster.

Traditionally the modifier algorithm determines the correction factor ΔP_i, based on a performance table [PRO 07]. The performance table is basically a two dimensional look-up table similar to the fuzzy control table. The table represents the desired transient behavior of the plant. Thus a performance table designed for one plant could easily be used for another as it only represent the desired behavior. A practical simplification of this is a penalty equation proposed by Jantzen [JAN 98b] to replace the performance table.

$$\Delta P = G_p(e_i + \tau \times ce_i) \times T_s \qquad\qquad [7.10]$$

The correction factor is a function of the *learning gain* G_p, the *desired time constant* τ, and the *sample period* Ts. The learning gain effects the rate of convergence. Too small and the SOFC will take a long time to converge if at all, too large and the system will become unstable rapidly. Jantzen proposes rule of thumb equations to govern the selection of τ and G_p. The desired time constant should be bounded by the plant time constant, τ_p, and dead time, T_p.

$$T_p \leq \tau \leq T_p + \tau_p \tag{7.11}$$

The original equation states that G_p should be chosen so that each correction factor is not greater than 1/5 of the maximum value in the fuzzy table. This prevents the fuzzy table from winding up and causing the system to become unstable. Based on experimentation it was concluded that a magnitude of 1/10 produced better results.

The SOFC was initially trained offline to assess the quality of the trained controller compared to the standard fuzzy controller. The training was carried out in MATLAB Simulink environment. The entire PAM system had to be modeled to enable *offline* training. Figure 7.22 shows the Simulink SOFC block diagram.

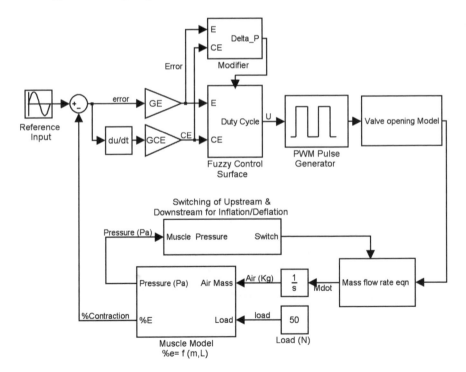

Figure 7.22. *Simulink SOFC diagram for offline tuning-position control*

In the final implementation an online training would be sufficient. The online training would negate the necessity to model any of the subsystems within the pneumatic system, and as such prove to be a more efficient solution when compared to model-based controllers.

The high-speed pneumatic valve model is included in the "valve-opening model" block. The two look-up tables to model the PAM are incorporated in the "muscle model" block and the "mass flow rate equation" block calculates the mass of air entering and leaving the PAM based on the valve opening and pressure difference (section 7.4).

7.5.4.1. *Simulation training results*

In order to properly train the SOFC offline, the reference input should match the reference input from the actual system; however, as this reference is to be derived from the exoskeleton user's surface EMG signal a simplified reference was used (i.e. a sinusoid signal).

Prior to implementing a force control system (as would be implemented in the final exoskeleton), a position control system was implemented as a preliminary test of the SOFC performance. The feedback in this simulation was the percentage contraction of the PAM and the reference was the intended percentage contraction.

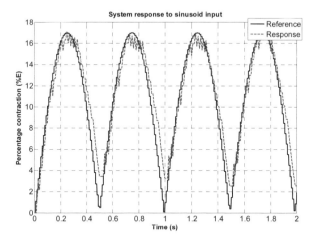

Figure 7.23. *Simulated PAM tracking response before training*

The SOFC was trained with a 2 Hz rectified sinusoid wave as this signal resembled the motion of the knee joint during normal walking movement. The frequency of 2 Hz was chosen as this was deemed to be the upper limit for the step frequency in a person requiring assistance (i.e. 2 steps per second per leg) [CHA 10]. The amplitude of the sine wave was set to 17% PAM contraction which is approximately 60° knee flexion. Figure 7.23 shows the PAM tracking before training and Figure 7.24 shows the tracking after a training session of 80 seconds.

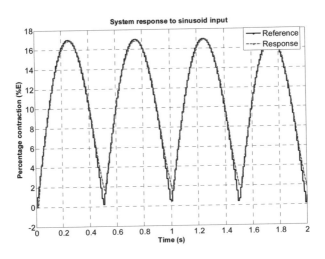

Figure 7.24. *Simulated PAM tracking response after training*

Simulation of the SOFC validates the ability of the modifier algorithm to improve the tracking accuracy of the fuzzy controller. The encouraging results obtained through simulation justify an experimental testing of the SOFC.

7.6. Surface electromyography

7.6.1. *Origins of surface electromyography (sEMG) signals*

The link between the user and the exoskeleton is achieved via surface myoelectric signals. These signals are intended to allow the user direct control of the exoskeleton and by this means provide intelligent control. The challenge however it to determine an exact mapping of the signals recorded on the skin surface to the muscle torque developed across the knee joint. This section is still part of an ongoing research and only preliminary findings are discussed. The sEMG origin, signal acquisition and processing, and sEMG to force mapping will be discussed in this chapter.

Muscles are both the dominant tissue and the primary organ of the human body. An estimated 70% to 85% of gross body weight is attributed to muscles [CRA 98].

Skeletal muscle, which is the type of muscle concerned here, produces torque across a joint by shortening its resting length. On a macroscopic level muscles are classified by their line of action, direction of pull and their origins[4] and insertions.[5] However closer investigation reveals that muscles are in fact composed of compartments. Thus instead of a unit of massive muscle, most muscles are made up of a series of smaller compartments that run along the same or different direction [CRA 98]. Muscles are surrounded by connective tissue that hold them together and prevent displacement from the line of action.

The functional unit of the neuromuscular system is the motor unit (MU). MUs consist of an α-motor neuron and the connected muscle fibers [STA 10]. The α-motor neurons, located in the spinal cord of the brain stem, creates an electrical impulse (action potential) that travels along axons to its terminal branches, each of which is connected to a single muscle fiber at the neuromuscular junction. The connection is usually in the middle or proximal to the middle. As the action potential (AP) reaches the muscle fibers, depolarization of the fiber membrane triggers muscle contraction. The membrane depolarization causes a time-varying transmembrane electric current field that can be measured non-invasively from the surface of the skin above the muscle [BAS 62].

Since a single action potential will only cause a twitch, to achieve a longer period of contraction a series of APs are generated by the motor neuron. The recruitment or firing of the muscle fibers is generally recognized to be random; however, studies have shown that smaller MUs located deep within the muscle are recruited first before larger MUs closer to the muscle surface are recruited [DAH 05].

The EMG signal detected on the skin surface is the algebraic summation of these APs [DAY 01]. As these waves are bi-phasic or tri-phasic, the phase cancelation that consequently occurs, results in the detected EMG signal having an amplitude less than proportional to the number of MU firing per second. The amplitude of the detected surface EMG also varies significantly as a result of the summation [KEE 08].

7.6.2. sEMG signal acquisition and conditioning

The muscle groups that are of interest for the present application are the knee extensor and flexor muscles. Three extensor and three flexor muscles were chosen based on their percentage cross-sectional area (%PCA) and the ability of signal detection using surface electrodes. The extensor muscles are vastus lateralis (20%),

---

4 The point where the muscle attaches to the bone that is closer to the center of the body.
5 The point where the muscle attaches to the bone that is furthest away from the center of the body.

rectus femoris (8%) and vastus medialis (15%). The flexor muscles are semi-tendinosus (3%), semi-membranosus (10%), and biceps femoris (10%) (Figure 7.25).

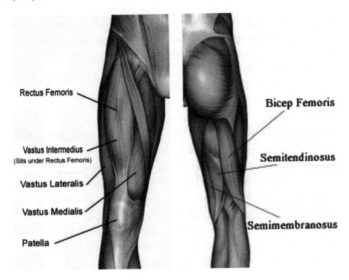

Figure 7.25. *Thigh extensor (left) and flexor (right) muscles that are monitored [BEE 07]*

Gold-plated bipolar electrodes were used in conjunction with a preconditioning circuit to preprocess the acquired sEMG signals. The preconditioning circuit layout is given in Figure 7.26. The instrumentation amplifier is essentially a high-input impedance differential amplifier with a high common mode rejection ratio. Next band pass filters (20–400 Hz) are utilized to section out the frequencies that have the most information (energy) regarding the muscle activation. A second amplifier is then used to further boost the signal to fall within the 0–5 V range. Thus the original signal is amplified 1,000 times.

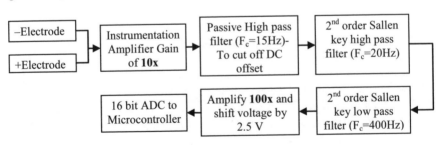

Figure 7.26. *Preamplifier circuit layout*

The preprocessed signal is sampled at 1,000 Hz, full wave rectified and then filtered using a moving average filter with a window of 100 values. Existing literature recommends low pass filtering at frequencies below 3 Hz to estimate the force envelope, however the phase delay (group delay) associated with classical filtering techniques is not desired in control applications. Physiologically there is a time delay between the onset of electrical activity and detection of force. This electromechanical delay (EMD) is estimated to range from 30 ms to 150 ms [STA 05], thus a moving average filter provides a constant delay (100 ms) within the EMD, which is not a function of the frequency (Figure 7.27 and 7.28).

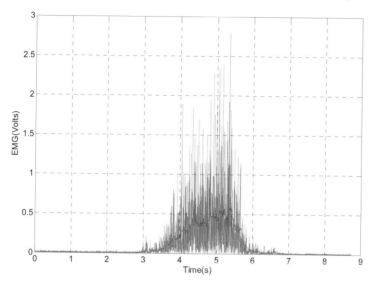

Figure 7.27. *Muscle activation pattern for rectus femoris during maximum voluntary isometric contraction at 75° knee angle. The red plot is the raw rectified signal and the blue plot is the rectified and filtered signal*

7.6.3. *Relating sEMG to muscle force*

The processed sEMG signals next have to be related to the force generated by each muscle [HAY 09, FLE 04]. Various methods have been proposed for this purpose, most of which are primarily concerned with the exact modeling of the sEMG to force relationship. These methods usually are not practical in a real-time control scheme. In the case of a lower limb exoskeleton a close approximation will suffice as the *assistance ratio* can be adjusted to provide a "comfortable" support.

Proportional myoelectric control was implemented by Ferris *et al.* [FER 06] to control an ankle–foot orthosis. The acquired sEMG signal was first rectified and low pass filtered, then used to proportionally control the pressure within PAMs. This linear and simplistic approximation to the muscle force was sufficient for the intended tests. In his PhD thesis Fleischer [FLE 07] proposed a simplified biochemical body model to relate the acquired EMG to the force produced. The model was based on the traditional Hill type model which approximates each muscle group to a contractile element that creates active force in parallel with a passive element that resists stretching through passive force. The model developed displayed impressive results; however, certain key parameters such as the force–velocity correlation and the change of the optimal muscle fiber length with respect to the activation level were left out to reduce the number of parameters that needed to be calibrated each time. These crucial parameters could in theory improve the accuracy of the model.

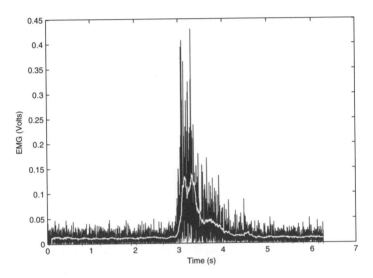

Figure 7.28. *EMG activation pattern during a single sit-to-stand movement (Black-Raw rectified EMG, White rectified and filtered EMG)*

In the current work, the preliminary control signals for the PAM actuators are derived from sEMG signals in a similar fashion to the method employed by Ferris *et al.*, that is proportional myoelectric control. This initial setup will be replaced by an artificial neural network (ANN) model that would be able to capture the nonlinear relation between the knee angle (θ), knee velocity ($\dot{\theta}$), the six sEMG signals, and the resultant torque across the joint. Similar works by Luh *et al.* [LUH 99] and Hahn [HAH 07] have demonstrated the viability and the effectiveness of such an approach.

7.7. Hardware implementation

A real-time fuzzy control system was implemented to verify the results obtained through simulation. National Instruments compact RIO (NIcRIO-9074) together with LabVIEW graphical programming language were used as the real time hardware. The facility of FPGA (Field Programmable Gate Array) allowed fast and deterministic input sampling and control signal output. The input into the control system is the position from the encoder (US Digital S5S-360 pulse/rev) and the outputs are the PWM signals to the two high-speed valves. The PAM used for the experiment is the MAS-10-N290 (10 mm internal diameter and 290 mm relaxed length) from Festo Inc (Figure 7.29).

The experiment was carried out for the PAM position (percentage contraction) control. As mentioned in section 7.5.4.1, position control was implemented as an initial effort to ascertain the performance of a SOFC. The trained fuzzy surface was implemented directly into the LabVIEW code as a 2D array. The entire control algorithm could be executed within 2 ms and since the PWM period is 10 ms (section 7.4) this was well within the allocated time period.

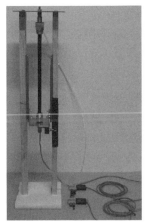

Figure 7.29. *Test rig for position control (left), the two high-speed valves are connected to a t-joint to enable PAM inflation and deflation. A closer view of the encoder (right)*

The tracking accuracy of the controller was tested with step inputs. The ideal behavior would be for the PAM to minimize air losses as this is undesirable if it is to be implemented in a mobile orthosis. Thus a certain amount of undershoot is preferred to overshoot.

Three test inputs were used, step input of 5%, 10%, and 15% contraction. A loading of 4.55 kg was attached to the lower end of the PAM, to represent the leg weight. The test was repeated three times and the average steady-state error was calculated. The steady-state error for each contraction level is given in Table 7.3.

Input	Steady-state error (%)
5% contraction with 4.5 kg load	1.2
10% contraction with 4.5 kg load	1.76
15% contraction with 4.5 kg load	1.25
10% contraction with no load	0.07

Table 7.3. *Experimental steady-state error for various input signals*

The unloaded system response for a given reference input is much smoother (Figure 7.30) than a system with loading. The main reason for this is the inertia of the load and the spring-like characteristic of the PAM. The accelerated load compresses the PAM and causes the system to oscillate slightly (Figure 7.31). This behavior is only observed during a step reference input; however for ramp or sinusoid input the tracking is smoother.

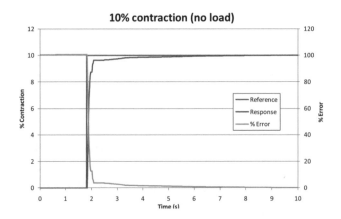

Figure 7.30. *System response to a 10% contraction input without loading*

The next step is to perform online training of the SOFC and to compare its performance with an offline-trained SOFC. If the online training proves to be successful then the assumption can be made that the offline training can be omitted when developing the force control system.

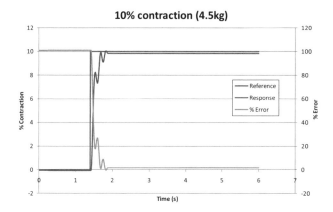

Figure 7.31. *System response to a 10% contraction input with loading*

7.8. Concluding remarks

The design simulation and experimental validation of a self-organizing fuzzy controller to achieve accurate position control of a PAM for a lower limb exoskeleton is discussed. Simulation and experimental results justify the application of a fuzzy controller to control the highly nonlinear PAM pneumatic system. Though classical model-based approaches have been used in the literature to achieve PAM control, the facility with which the fuzzy controller is able to achieve comparable results is impressive.

When integrated with self-organizing capability the controller is able to quickly adapt the control surface to suit the controlled system. It is envisaged that the fuzzy force control system that will ultimately be implemented in the exoskeleton, will obtain similar performance results as those from the position control.

Similarly, it is hypothesized that artificial intelligence methods employed to map the nonlinear relationship between the force developed by the knee muscles (extensor and flexor) and the measured EMG signals will outperform model-based approaches in terms of simplicity. Neural networks are generally recognized to be an excellent universal approximator, thus this characteristic can be exploited to model the sEMG–force relationship.

In conclusion, the application of fuzzy logic and neural networks, along with a compliant and skeletal muscle-like actuator, is deemed to be an ideal combination for a powered lower limb exoskeleton.

7.9. Acknowledgment

The authors would like to acknowledge Mr Markus Dreher and Festo Inc. for providing the fluidic muscles (PAM) used in this research. The advice provided by Mr Campbell Lintott (Festo NZ) and Julian Murphy (University of Canterbury) is also acknowledged.

7.10. Bibliography

[ANH 08] ANH H.P.H., AHN K.K., IL YOON J., ICROS/KOMMA, "Identification of the 2-axes pneumatic artificial muscle (PAM) robot arm using double NARX fuzzy model and genetic algorithm", *International Conference on Smart Manufacturing Application*, Goyangsi, South Korea, 9–11 April 2008.

[BAL 03] BALASUBRAMANIAN K., RATTAN K.S., "Fuzzy logic control of a pneumatic muscle system using a linearizing control scheme", *22nd International Conference of the North-American-Fuzzy-Information-Processing-Society* (NAFIPS) Chicago, Il, 24–26 July 2008.

[BAR 97] BARBER A., 1997, "Pipe flow calculations", in *Pneumatic Handbook*, edited by E. S. Ltd, Oxford, Elsevier Advance Technology.

[BAS 62] BASMAJIAN J.V., "Muscles alive: Their functions revealed by electromyography", *Journal of Medical Education*, vol. 37, no. 8, 1962, p. 802.

[BEE 07] WWW.BEEBLEBLOG.COM, "Diagram of hamstring muscles", edited by B. s. F. Blog, 2007.

[BON 09] BONG-SOO K., KOTHERA C.S., WOODS B.K.S., WERELEY N.M., "Dynamic modeling of Mckibben pneumatic artificial muscles for antagonistic actuation", *IEEE International Conference on Robotics and Automation*, 2009.

[CAL 95] CALDWELL D.G., MEDRANOCERDA G.A., GOODWIN M., "Control of pneumatic muscle actuators", *IEEE Control Systems Magazine*, vol. 15, no. 1, 1995, pp. 40–48.

[CHA 10] CHANDRAPAL M., CHEN X., WANG W., "Self-organizing fuzzy control of pneumatic artificial muscle for acitve orthotic device", *6th IEEE Conference on Automation Science and Engineering*, Toronto, Canada, August 2010.

[CHE 07] CHEN Y., ZHANG J.F., YANG C.J., NIU B., "Design and hybrid control of the pneumatic force-feedback systems for arm-exoskeleton by using on/off valve", *Mechatronics*, vol. 17, no. 6, 2007, pp. 325–335.

[CHO 96] CHOU C.P., HANNAFORD B., "Measurement and modeling of McKibben pneumatic artificial muscles", *IEEE Transactions on Robotics and Automation*", vol. 12, no. 1, 1996, pp. 90–102.

[CHO 06] CHOI T., LEE J., LEE J., "Control of artificial pneumatic muscle for robot application", *Proceedings of the 2006 IEEE/RSJ International Conference on Intelligent Robots and Systems*, Beijing, China, 9–15 October 2006.

[CRA 98] CRAM J.R., KASMAN G.S., HOLTZ J., *Introduction to Surface Electromyography*, 1st edition, Gaithersburg, Maryland, Aspen Publication, 1998.

[DAE 02] DAERDEN F., LEFEBER D., "Pneumatic artificial muscles: Actuators for robotics and automation", *European Journal of Mechanical and Environmental Engineering*, vol. 47, 2002, p.10.

[DAH 05] DAHMANE R., DJORDJEVIC S., SIMUNIC B., VALENCIC V., "Spatial fiber type distribution in normal human muscle: Histochemical and tensiomyographical evaluation", *Journal of Biomechanics*, vol. 38, no. 12, 2005, pp. 2451–2459.

[DAY 01] DAY S.J., HULLIGER M., "Experimental simulation of cat electromyogram: Evidence for algebraic summation of motor-unit action-potential trains", *Journal of Neurophysiology*, vol. 86, no. 5, 2001, pp. 2144–2158.

[FER 05] FERRIS D.P., CZERNIECKI J.M., HANNAFORD B., "An ankle-foot orthosis powered by artificial pneumatic muscles", *Journal of Applied Biomechanics*, vol. 21, no. 2, 2005, pp. 189–197.

[FER 06] FERRIS D.P., GORDON K.E., SAWICKI G.S., PEETHAMBARAN A., "An improved powered ankle-foot orthosis using proportional myoelectric control", *Gait Posture*, vol. 23, no. 4, 2006, pp. 425–428.

[FES 03] FESTO A.G., CO KG, "Fluidic muscle DMSP, with press-fitted connections & fluidic muscle MAS, with screwed connections", *Info 501*, edited by F. A. C. KG, Festo AG & Co. KG, 2003.

[FES 10] FESTO A.C., 2010, "Fluidic muscle", Festo AG & Co., 2010 [cited 14 December 2010]. Available http://www.festo.com/rep/el_gr/assets/Corporate_img/Fluidicmuscle_0375mu_500px.jpg.

[FLE 04] FLEISCHER C., KONDAK K., REINICKE C., HOMMEL G., "Online calibration of the EMG to force relationship", *Proceedings of the 2004 IEEE/RSJ International Conference on Intelligent Robots and Systems* (IROS 2004), 28 September–2 October, 2004.

[FLE 07] FLEISCHER C., "Controlling exoskeletons with EMG signals and a biomechanical body model", *Faculty IV – Electrical Engineering and Computer Science*, Technical University of Berlin, Berlin, 2007.

[GOR 06] GORDON K.E., SAWICKI G.S., FERRIS D.P., "Mechanical performance of artificial pneumatic muscles to power an ankle-foot orthosis", *Journal of Biomechanics*, vol. 39, no. 10, 2006, pp. 1832–1841.

[HAH 07] HAHN M.E., "Feasibility of estimating isokinetic knee torque using a neural network model", *Journal of Biomechanics*, vol. 40, no. 5, 2007, pp. 1107–1114.

[HAY 05] HAYASHI T., KAWAMOTO H., SANKAI Y., "Control method of robot suit HAL working as operator's muscle using biological and dynamical information", *Proceedings of the 2005 IEEE/RSJ International Conference on Intelligent Robots and Systems*, (IROS 2005), 2005.

[HAY 09] HAYASHIBE M., GUIRAUD D., POIGNET P., "EMG-to-force estimation with full-scale physiology based muscle model", *Proceedings of the 2009 IEEE/RSJ International Conference on Intelligent Robots and Systems* (IROS 2009), 10–15 October 2009.

[HOO 09] HOOTMAN J.H.C., THEIS KA, BRAULT MW, ARMOUR BS, "Prevalence and most common causes of disability among adults – United States, 2005", *Morbidity and Mortality Weekly Report*, MMWR, vol. 58, no. 16, 2009, pp. 421–426.

[HUA 05] HUANG L.H., STEGER J.R., KAZEROONI H., "Hybrid control of the berkeley lower extremity exoskeleton (BLEEX)", *ASME International Mechanical Engineering Congress and Exposition*, Orlando, FL, 5–11 November 2005.

[INO 88] INOUE K., "Rubbertuators and applications for robots", *The 4th International Symposium on Robotics Research*, Cambridge, MA, 1988.

[JAN 98] JANTZEN J., Design of fuzzy controllers, DK-2800 Lyngby, Denmark, Technical University of Denmark, Department of Automation, 1998.

[JAN 98b] JANTZEN J., The self-organising fuzzy controller, DK-2800 Lyngby, Denmark, Technical University of Denmark, Department of Automation, 1998.

[JAN 98c] JANTZEN J., Tutorial on fuzzy logic, DK-2800 Lyngby, Denmark, Technical University of Denmark, Department of Automation, 1998.

[KAJ 95] KAJIMA T., KAWAMURA Y., "Development of a high-speed solenoid valve – investigation of solenoids", *IEEE Transactions on Industrial Electronics*, vol. 42, no. 1, 1995, pp. 1–8.

[KAW 03a] KAWAMOTO H., KANBE S., SANKAI Y., "Power assist method for HAL-3 estimating operator's intention based on motion information", *Proceedings of the 12th IEEE International Workshop on Robot and Human Interactive Communication* (ROMAN 2003), 2003.

[KAW 03b] KAWAMOTO H., SUWOONG L., KANBE S., SANKAI Y., "Power assist method for HAL-3 using EMG-based feedback controller", *IEEE International Conference on Systems, Man and Cybernetics*, 2003.

[KAW 04] KAWAMOTO H., SANKAI Y., "Power assist method based on phase sequence driven by interaction between human and robot suit", *Proceedings of the 13th IEEE International Workshop on Robot and Human Interactive Communication* (ROMAN 2004), 2004.

[KAW 09] KAWAMOTO H., HAYASHI T., SAKURAI T., EGUCHI K., SANKAI Y., "Development of single leg version of HAL for hemiplegia", *Annual International Conference of the IEEE on Engineering in Medicine and Biology Society* (EMBC 2009), 2009.

[KAZ 05] KAZEROONI H., RACINE J.L., HUANG L.H., STEGER R., "On the control of the Berkeley Lower Extremity Exoskeleton (BLEEX)", *IEEE International Conference on Robotics and Automation* (ICRA), Barcelona, Spain, 18–22 April 2005.

[KEE 08] KEENAN K.G., VALERO-CUEVAS F.J., "Epoch length to accurately estimate the amplitude of interference EMG is likely the result of unavoidable amplitude cancellation", *Biomedical Signal Processing and Control*, vol. 3, no. 2, 2008, pp. 154–162.

[KOV 06] KOVAČIĆ Z., BOGDAN S., *Fuzzy Controller Design: Theory and Applications – Technology & Engineering*, CRC /Taylor & Francis, 2006.

[KRE 06] KREBS H.I., HOGAN N., "Therapeutic robotics: a technology push", *Proceedings of the IEEE*, vol. 94, no. 9, 2006, pp.1727–1738.

[LIL 03] LILLY J.H., "Adaptive tracking for pneumatic muscle actuators in bicep and tricep configurations", *IEEE Transactions on Neural Systems and Rehabilitation Engineering*, vol. 11, no. 3, 2003, pp. 333–339.

[LLO 10] LLOYD-JONES D., ADAMS R.J., BROWN T.M., CARNETHON M., DAI S., DE SIMONE G., FERGUSON T.B., FORD E., FURIE K., GILLESPIE C., GO A., GREENLUND K., HAASE N., HAILPERN S., HO P.M., HOWARD V., KISSELA B., KITTNER S., LACKLAND D., LISABETH L., MARELLI A., MCDERMOTT M.M., MEIGS J., MOZAFFARIAN D., MUSSOLINO M., NICHOL G., ROGER V.L., ROSAMOND W., SACCO R., SORLIE P., THOM T., WASSERTHIEL-SMOLLER S., WONG N.D., WYLIE-ROSETT J., "Heart disease and stroke statistics – 2010 update: A report from the American Heart Association", *Circulation*, vol. 121, no. 7, 2010, pp. e46–e215.

[LOW 06] LOW K.H., YIN Y.H., "Providing assistance to knee in the design of a portable active orthotic device", *IEEE International Conference on Automation Science and Engineering*, Shanghai, China, 08–10 October 2006.

[LUC 09] LUCAS K., "The 'all or none' contraction of the amphibian skeletal muscle fiber", *The Journal of Physiology*, vol. 38, no. 2–3, 1909, pp. 113–133.

[LUH 99] LUH J.-J., CHANG G.-C., CHENG C.-K., LAI J.-S., KUO T.-S., "Isokinetic elbow joint torques estimation from surface EMG and joint kinematic data: Using an artificial neural network model", *Journal of Electromyography and Kinesiology*, vol. 9, no. 3, 1999, pp. 173–183.

[MAM 75] MAMDANI E.H., BAAKLINI N., "Prescriptive method for deriving control policy in a fuzzy-logic controller", *Electronics Letters*, vol. 11, no. 25, 1975, pp. 625–626.

[MIN 97] MING-CHANG S., CHUEN-GUEY H., "Fuzzy PWM control of the positions of a pneumatic robot cylinder using high speed solenoid valve", *JSME International Journal*, vol. 40, no. 3, 1997, pp. 469–476.

[MIN 10] MINH T.V., TJAHJOWIDODO T., RAMON H., VAN BRUSSEL H., "Cascade position control of a single pneumatic artificial muscle–mass system with hysteresis compensation", *Mechatronics*, vol. 20, no. 3, 2010, pp. 402–414.

[NIK 06] NIKITCZUK J., DAS A., VYAS H., WEINBERG B., MAVROIDIS C., "Adaptive torque control of electro-rheological fluid brakes used in active knee rehabilitation devices", *Proceedings of the 2006 IEEE International Conference on Robotics and Automation* (ICRA 2006), 2006.

[OHN 06] OHNABE H., "Current trends in rehabilitation engineering in Japan", *Assistive Technology*, vol. 18, no. 2, 2006, pp. 220–232.

[PAC 97] PACK R.T., CHRISTOPHER J.L., KAWAMURA K., "A rubbertuator-based structure-climbing inspection robot", *Proceedings of the IEEE International Conference on Robotics and Automation*, Albuquerque, New Mexico,1997.

[POT 04] POTVIN J.R., BROWN S.H.M., "Less is more: high pass filtering, to remove up to 99% of the surface EMG signal power, improves EMG-based biceps brachii muscle force estimates", *Journal of Electromyography and Kinesiology*, vol. 14, no. 3, 2004, pp. 389–399.

[PRA 04] PRATT J.E., KRUPP B.T., MORSE C.J., COLLINS S.H., "The robo knee: An exoskeleton for enhancing strength and endurance during walking", *IEEE International Conference on Robotics and Automation*, New Orleans, LA, 26 April–01 May 2004.

[PRO 07] PROCYK T.J., MAMDANI E.H., "A linguistic self-organizing process controller", *Automatica*, vol. 15, no. 1, 1979, pp. 15–30.

[SAW 05] SAWICKI G.S., GORDON K.E., FERRIS D.P., "Powered lower limb orthoses: Applications in motor adaptation and rehabilitation", *9th IEEE International Conference on Rehabilitation Robotics*, Chicago, IL, 28 June–01 July 2005.

[SAW 08] SAWICKI G.S., FERRIS D.P., "Mechanics and energetics of level walking with powered ankle exoskeletons", *Journal of Experimental Biology*, vol. 211, no. 9, 2008, pp. 1402–1413.

[SAW 09] SAWICKI G.S., "A pneumatically powered knee-ankle-foot orthosis (KAFO) with myoelectric activation and inhibition", *Journal of Neuroengineering and Rehabilitation*, vol. 6, 2009, p. 16.

[SCH 61] SCHULTE H.F., "The characteristics of the McKibben artificial muscle", *The Application of External Power in Prosthetics and Orthotics*, Appendix H, vol. 87, 1961, pp. 94–115.

[SEN 09] SENANAYAKE C., SENANAYAKE S.M.N.A., "Emerging robotics devices for therapeutic rehabilitation of the lower extremity", *IEEE/ASME International Conference on Advanced Intelligent Mechatronics* (AIM 2009), 2009.

[SHI 02] SHIGERU Y., "Assistive engineering: a new engineering discipline", *Journal of the Japan Society of Mechanical Engineers*, vol. 105, no. 1002, 2002, pp. 315–317.

[SIT 08] SITUM Z., HERCEG S., "Design and control of a manipulator arm driven by pneumatic muscle actuators", *16th Mediterranean Conference on Control and Automation*, Ajaccio, France, 25–27 June 2008.

[STA 05] STAUDENMANN D., KINGMA I., STEGEMAN D.F., VAN DIEEN J.H., "Towards optimal multi-channel EMG electrode configurations in muscle force estimation: A high density EMG study", *Journal of Electromyography and Kinesiology*, vol. 15, no. 1, 2005, pp. 1–11.

[STA 10a] STATISTICS BUREAU OF JAPAN, "Population", *Statistical Handbook of Japan 2010*, edited by Statistics Bureau.

[STA 10b] STAUDENMANN D., ROELEVELD K., STEGEMAN D.F., VAN DIEEN J.H., "Methodological aspects of SEMG recordings for force estimation – a tutorial and review", *Journal of Electromyography and Kinesiology*, vol. 20, no. 3, 2010, pp. 375–387.

[TAK 06] TAKUMA T., HOSODA K., "Controlling the walking period of a pneumatic muscle walker", *International Journal of Robotics Research*, vol. 25, no. 9, 2006, pp. 861–866.

[TON 00] TONDU B., LOPEZ P., "Modelling and control of McKibben artificial muscle robot actuators", *IEEE Control Systems Magazine*, vol. 20, no. 2, 2000, pp. 15–38.

[UBE 07] WWW.UBERGIZMO.COM, 2010, Exoskeleton up for rent 2007 (cited 14 December 2010).

[VUK 90] VUKOBRATOVIC M., BOROVAC B., SURLA D., STOKIC D., *Biped Locomotion: Dynamics, Stability, Control and Application*, Springer-Verlag, 1990.

[WEI 07] WEINBERG B., NIKITCZUK J., PATEL S., PATRITTI B., MAVROIDIS C., BONATO P., CANAVAN P., "Design, control and human testing of an active knee rehabilitation orthotic device", *IEEE International Conference on Robotics and Automation*, Rome, Italy, 10–11 April 2007.

[XIA 08] XIAO L., HONG X., TING G., "Development of legs rehabilitation exercise system driven by pneumatic muscle actuator", *The 2nd International Conference on Bioinformatics and Biomedical Engineering* (ICBBE 2008), 2008.

[ZAD 65] ZADEH L.A., "Fuzzy sets", *Information and Control*, vol. 8, no. 3, 1965, pp. 338–353.

[ZAD 68] ZADEH L.A., "Fuzzy algorithms", *Information and Control,* vol. 12, no. 2, 1968, pp. 94–102.

[ZHA 08] ZHANG J.-F., YANG C.-J., CHEN Y., ZHANG Y., DONG Y.-M., "Modeling and control of a curved pneumatic muscle actuator for wearable elbow exoskeleton", *Mechatronics*, vol. 18, no. 8, 2008, pp. 448–457.

List of Authors

Bilin AKSUN GÜVENÇ
Department of Mechanical
Engineering
School of Engineering and
Architecture
Okan University Tuzla Campus
İstanbul
Turkey

Kean AW
Department of Mechanical
Engineering
University of Auckland
New Zealand

Mervin CHANDRAPAL
Mechatronics Research Laboratory
Department of Mechanical
Engineering
University of Canterbury
New Zealand

Xiaoqi CHEN
Mechatronics Research Laboratory
Department of Mechanical
Engineering
University of Canterbury
New Zealand

Paulo J. DAVIM
Department of Mechanical
Engineering
University of Aveiro
Campus Santiago
Portugal

Burak DEMİREL
School of Electrical Engineering
Automatic Control
KTH Royal Institute of Technology
Stockholm
Sweden

João M. G. FIGUEIREDO
CEM-Centro de Engenharia
Mecatrónica/IDMEC-IST
University of Évora
Portugal

Levent GÜVENÇ
Department of Mechanical
Engineering
School of Engineering and
Architecture
Okan University Tuzla Campus
İstanbul
Turkey

Maki K. HABIB
Mechanical Engineering
Department
School of Sciences and Engineering
American University in Cairo
Egypt

Shahid HUSSAIN
Department of Mechanical
Engineering
University of Auckland
New Zealand

Yukihiro KUSUMOTO
Interior & Design Research Institute
Fukuoka Industrial Technology
Center
Ohkawa
Japan

Andrew MCDAID
Department of Mechanical
Engineering
University of Auckland
New Zealand

Fusaomi NAGATA
Department of Mechanical
Engineering
Tokyo University of Science
Sanyo-Onoda
Japan

Serkan NECİPOĞLU
Mekar Mechatronics Research Labs
Department of Mechanical
Engineering
Istanbul Technical University
Istanbul
Turkey

Hiroo WAKAUMI
Tokyo Metropolitan College of
Industrial Technology
Japan

Wenhui WANG
Mechatronics Research Laboratory
Department of Mechanical
Engineering
University of Canterbury
New Zealand

Keigo WATANABE
Department of Intelligent Mechanical
Systems
Graduate School of Natural Science
and Technology
Okayama University
Japan

Sheng Q. XIE
Department of Mechanical
Engineering
University of Auckland
New Zealand

Index

A

active orthosis, 169, 171-174, 178-185, 187-189
AFM system model, 106
analog angular sensors, 69-99
assistive devices, 195-197
assistive/rehabilitative (AR), 197, 202
atomic force microscopy (AFM), 103-107, 115, 117-119, 128-129
automated identification, 133, 134, 164

B

biomechanics of gait, 172

C

Cartesian-type robot, 1, 21-23, 26
computer-aided design/computer-aided manufacturing (CAD/CAM), 1, 2, 8, 11, 16, 17, 19-20, 26, 27
control of mechatronics systems (COMES), 104-105, 111-113, 128
control system, 10, 22-25, 27, 29, 33, 50-52, 54-55, 58, 64
controller parameter space, 110

correction tools, 94-95, 99

D

decoding for identification, 139
design criteria, 179
dual-threshold method, 150

E

electromagnetic induction technology, 161
electromechanical coupling, 34, 38
electromechanical IPMC model, 33-34
envelope differential composite method, 155-156, 164
experimental validation, 46

F

finishing experiment, 25-26
five-axis NC machine tool, 1-5
fixed-period delay method, 158, 160, 164
force feedback loop, 11-12, 20, 22-23
frequency characteristic, 24-25
fuzzy control, 214-216, 219, 221, 229